Chronobioengineering

Introduction to Biological Rhythms with Applications, Volume 1

Synthesis Lectures on Biomedical Engineering

Editor
John D. Enderle, *University of Connecticut*

Lectures in Biomedical Engineering will be comprised of 75- to 150-page publications on advanced and state-of-the-art topics that spans the field of biomedical engineering, from the atom and molecule to large diagnostic equipment. Each lecture covers, for that topic, the fundamental principles in a unified manner, develops underlying concepts needed for sequential material, and progresses to more advanced topics. Computer software and multimedia, when appropriate and available, is included for simulation, computation, visualization and design. The authors selected to write the lectures are leading experts on the subject who have extensive background in theory, application and design.

The series is designed to meet the demands of the 21st century technology and the rapid advancements in the all-encompassing field of biomedical engineering that includes biochemical, biomaterials, biomechanics, bioinstrumentation, physiological modeling, biosignal processing, bioinformatics, biocomplexity, medical and molecular imaging, rehabilitation engineering, biomimetic nano-electrokinetics, biosensors, biotechnology, clinical engineering, biomedical devices, drug discovery and delivery systems, tissue engineering, proteomics, functional genomics, molecular and cellular engineering.

Chronobioengineering: Introduction to Biological Rhythms with Applications, Volume 1
Donald McEachron
2012

Capstone Design Courses, Part II: Preparing Biomedical Engineers for the Real World
Jay R. Goldberg
2012

Ethics for Bioengineers
Monique Frize
2011

Computational Genomic Signatures
Ozkan Ufuk Nalbantoglu and Khalid Sayood
2011

Advanced Probability Theory for Biomedical Engineers
John D. Enderle, David C. Farden, and Daniel J. Krause
2006

Intermediate Probability Theory for Biomedical Engineers
John D. Enderle, David C. Farden, and Daniel J. Krause
2006

Basic Probability Theory for Biomedical Engineers
John D. Enderle, David C. Farden, and Daniel J. Krause
2006

Sensory Organ Replacement and Repair
Gerald E. Miller
2006

Artificial Organs
Gerald E. Miller
2006

Signal Processing of Random Physiological Signals
Charles S. Lessard
2006

Image and Signal Processing for Networked E-Health Applications
Ilias G. Maglogiannis, Kostas Karpouzis, and Manolis Wallace
2006

Chronobioengineering: Introduction to Biological Rhythms with Applications, Volume 1
Donald McEachron

ISBN: 978-3-031-00525-1   paperback
ISBN: 978-3-031-01653-0   ebook

DOI 10.1007/978-3-031-01653-0

A Publication in the Springer series
SYNTHESIS LECTURES ON BIOMEDICAL ENGINEERING

Lecture #43
Series Editor: John D. Enderle, *University of Connecticut*
Series ISSN
Synthesis Lectures on Biomedical Engineering
Print 1930-0328   Electronic 1930-0336

Chronobioengineering

Introduction to Biological Rhythms with Applications, Volume 1

Donald McEachron
Drexel University

SYNTHESIS LECTURES ON BIOMEDICAL ENGINEERING #43

ABSTRACT

This book represents the first in a two-volume set on biological rhythms. This volume focuses on supporting the claim that biological rhythms are universal and essential characteristics of living organisms, critical for proper functioning of any living system.

The author begins by examining the potential reasons for the evolution of biological rhythms: (1) the need for complex, goal-oriented devices to control the timing of their activities; (2) the inherent tendency of feedback control systems to oscillate; and (3) the existence of stable and powerful geophysical cycles to which all organisms must adapt. To investigate the second reason, the author enlists the help of biomedical engineering students to develop mathematical models of various biological systems. One such model involves a typical endocrine feedback system. By adjusting various model parameters, it was found that creating a oscillation in any component of the model generated a rhythmic cascade that made the entire system oscillate. This same approach was used to show how daily light/dark cycles could cascade rhythmic patterns throughout ecosystems and within organisms.

Following up on these results, the author discusses how the twin requirements of *internal synchronization* (precise temporal order necessary for the proper functioning of organisms as complex, goal-oriented devices) and *external synchronization* (aligning organisms' behavior and physiology with geophysical cycles) supported the evolution of biological clocks. The author then investigates the clock systems that evolved using both conceptual and mathematical models, with the assistance of Dr. Bahrad Sokhansanj, who contributes a chapter on mathematical formulations and models of rhythmic phenomena. With the ubiquity of biological rhythms established, the author suggests a new classification system—the F^4LM approach (Function; Frequency; waveForm; Flexibility; Level of biological system expressing rhythms; and Mode of rhythm generation) to investigate biological rhythms. This approach is first used on the more familiar cardiac cycle and then on neural rhythms as exemplified and measured by the electroencephalogram. During the process of investigating neural cycles, the author finds yet another reason for the evolution of biological rhythms—physical constraints, such as those imposed upon long distance neural signaling. In addition, a common theme emerges of a select number of autorhythmic biological oscillators imposing coherent rhythmicity on a larger network or system. During the course of the volume, the author uses a variety of observations, models, experimental results, and arguments to support the original claim of the importance and universality of biological rhythms. In Volume 2, the author will move from the establishment of the critical nature of biological rhythms to how these phenomena may be used to improve human health, well-being, and productivity. In a sense, Volume 1 focuses on the *chronobio* aspect of chronobioengineering while Volume 2 investigates methods of translating this knowledge into applications, the *engineering* aspect of chronobioengineering.

KEYWORDS

biological rhythms, biological clocks, chronobiology, neural oscillators, physiological rhythms, mathematical modeling, feedback oscillation

This book is dedicated to the memory of Dr. Elisabeth Papazoglou, an exceptional researcher, teacher, and mentor, a colleague at the School of Biomedical Engineering, Science, and Health Systems, Drexel University, who enriched everyone she met. She was a true hero to us all; a life, although far too short, that continues to inspire.

Contents

Preface

When my colleague at Drexel University, Jay Bhatt, suggested this book, I approached the task with a mixture of enthusiasm and concern. I enjoy writing and the idea that someone—anyone—wanted me to write a text was flattering to say the least. However, even though I had been teaching a two-quarter course sequence on chronobiology for many years at Drexel and had taught similar courses at the University of Pennsylvania as well, I knew I was not a top name in the field. It had been many years since I had worked actively in research and even more since my Ph.D. in Neuroscience under the mentorship of Dr. Daniel Kripke at the University of California at San Diego (just how many years, I am unwilling to divulge). What was I thinking to suppose there was anything I could contribute to the teaching of chronobiology?

Despite these misgivings, I decided to begin the project. In so doing, I had to decide what approach to take. Certainly, there are already many excellent books on chronobiology (William Koukkari and Robert Sothern's *Introducing Biological Rhythms* and Jay Dunlap, Jennifer Loros and Patricia DeCoursey's *Chronobiology: Biological Timekeeping* come to mind among several others) so how could I differentiate myself from these admirable volumes? Several factors played into the decision to arrange the book as I have done.

First, the primary audience was to be biomedical engineers, although I certainly hope to appeal to many others as well. Thus, I needed to use my years of experience in teaching biomedical engineers to design the text to appeal to their specific mindset and points of view. What kind of mindset do biomedical engineering students have to which I must appeal? Primarily, the students display a design mindset. They are, after all, engineers and they approach information with applications in mind. Information by itself is never enough, the important question is what can be done with it. What products can be designed, and what processes can be developed to take advantage of this information? This led to two approaches, one used in this volume and one developed in Volume 2 (I will discuss why there are two volumes instead of one in a bit). I decided that I should approach biological rhythms in terms of design questions—why would biological rhythms evolve? What purposes to they serve? How do living organisms use biological rhythms? In this way, I hope to tap into the natural way engineers tend to approach information in order to ensure the relevance of that information.

Relevance is another factor. Despite all the research over the years accomplished by the many exceptionally talented individuals—Franz Halberg, Jurgen Aschoff, Colin Pittendrigh, Serge Daan, Charles Czeisler, David Dinges, Francis Levi and many others—there is a lack of appreciation for the need to understand chronobiology and biological rhythms as a critical part of all living systems. Most biology and physiology textbooks give biological rhythms only a cursory examination and students taught physiology seldom understand the dynamic nature of biochemistry, physiology and behavior. In fact, many images in standard physiology texts portray an incorrect static impression of inherently oscillatory activities. Over my years of teaching, I am constantly amazed when speaking with professionals in health care how little chronobiology is recognized for its impact on prevention and treatment of disease. Nor is this limited to health care. Several years ago, I was presenting data on the impact of time-of-day on the activities of pharmaceutical agents and a biologist in the audience raised his hand to claim that the data were not possible. Despite having no actual argument with how the experiments were conducted, the individual was simply unable to alter his mindset to accept what was clearly in front of him. This reluctance to incorporate the temporal dimension into an understanding of biological systems led me to take an advocacy position in the text. Thus, this text is not simply a compilation of data and information, it is a series of claims, arguments and assertions supported by evidence to demonstrate the importance and relevance of biological rhythms in understanding living systems.

The latter also reflects my experience in teaching over the past 28 years. During that time, and for various reasons, students have become increasing passive in their approaches to learning. They expect to have the answers laid out in front of them and be told what is right and wrong. They then memorize these data and regurgitate it back on examinations. Instructors often feed into this expectation by presenting scientific hypotheses and evidence as if they were facts simply to be accepted. As a result, students have adopted what one described to me as the "final and flush" approach to learning. They memorize what is needed to get a good grade on examinations and other assignments but little attempt is made to incorporate new information into their pre-existing world views. Learning is also highly compartmentalized. After obtaining poor results on midterm in a class one year, I asked the students how it was that they did so badly given that the midterm was based upon material previously presented in several other courses. One student came to me after class and suggested that the reason was that I had asked questions related to chemical thermodynamics in a class on nervous systems. I should not expect students to integrate such information.

The observations reinforced my decision to create a text in the form of questions, claims, models, arguments, and evidence. I wanted to emphasize that science is a process, not an end point, and encourage readers to consider the questions posed and evaluate the answers suggested. I also wanted to create a text that was modular and mostly self-contained. Thus, there are discussions on important matters—the nature of scientific evidence, the use of evolution in biomedical research—not directly involved in the discussion of biological rhythms. Another reason for this approach is to emphasize the transdisciplinary nature of biomedical research. One cannot compartmentalize living systems and it is important to emphasize the inter-relatedness of all aspects of biology. This may annoy some of the more sophisticated readers who feel that they do not need to review such elementary matters. My apologies for this—no insult was intended—but I find that students do need such review and this text is designed as an educational exercise above all.

Finally, there is the question of two volumes. I could point out that I teach a two-quarter course in chronobioengineering at Drexel University and that this two-volume approach reflects that two-quarter course sequence. However, that would not really be accurate. The truth is that writing this text became a journey for me every bit the same way I intended it to be for the readers. The more research I conducted for the book, the more fun I had simply doing the research and the book began to get longer and longer. In addition, I could never convince myself that the work was really ready to be published. There was also some additional observation, some new experiment to be analyzed and incorporated. Finally, my guilt over the many delays prompted me to suggest a two-volume approach. The first volume, the one you are now reading, covers the basic arguments, models, and observations supported the importance of biological rhythms to the understanding of biology. Volume 2 will start from that understanding and investigate the many levels of biological rhythms and how this knowledge can be put to practical use.

As I indicated in the Acknowledgements, many individuals helped in the writing of this text. Many of the best sections came from their ideas and suggestions. Any errors in fact or argument are, however, mine and mine alone.

Donald McEachron
September 28, 2012

How to Use this Book

This book was written based upon two quarter-long courses taught at the School of Biomedical Engineering, Science and Health Systems at Drexel University in Philadelphia, PA, USA. These courses are taught to both undergraduate and graduate students as part of concentrations in biomechanics and human performance engineering and neuroengineering. Thus, the main target audience for the text is biomedical engineering students.

Despite this, I hope this volume will prove accessible to a far broader audience. With the exception of Chapter 6 written by Dr. Bahrad Sokhansanj, the background knowledge required to understand the material is not extensive or beyond that of an educated layperson. The perspective may be that of an engineer, but one does not need to be an engineer to read this text.

To fully appreciate Chapter 6 does require training through at least basic calculus. However, for those without such training, it is still possible to get much from the chapter by way of the discussion and figures. Many of the implications of the equations and models are described in sufficient detail so as to provide clarity and support for ideas presented elsewhere and thus well worth reading even for readers with a limited background in mathematics.

The chapters should be read in order insofar as the entire text represents an argument in support of the main proposition that biological rhythms are a ubiquitous and fundamental characteristic of biological systems. Each chapter represents a sequential step in this developing line of reasoning and thus each builds on the previous one. This is a journey where one should not try and get to the destination before making the trip.

The current climate in higher education is that of assessment, where instructional materials are expected to support specific student learning outcomes or goals. For undergraduate engineering, the premier accrediting body is ABET, Inc. and this organization promotes the use of very specific learning outcomes for engineers. In those terms, this text will help students with the following learning outcomes:

1. (ABET a) Ability to apply knowledge of mathematics, science, and engineering to solve problems at the interface of engineering and biology;

2. (ABET j) Knowledge of contemporary issues;

3. Knowledge of interdisciplinary concepts within a biomedical perspective.

Acknowledgments

No work of this kind is possible without the assistance of many people. I would like to acknowledge some them without whom this book would not have been possible. First, Dr. Bahrad Sokhansanj, who not only contributed Chapter 6 but also reviewed the other chapters and provided valuable advice and insight. Mr. Maurice Baynard also reviewed each chapter several times and his comments were invaluable in providing the final text. I am grateful to Rajarshi Ganguly and George Neusch who assisted the neuron modeling and especially Kevin Freedman, who converted my simple conceptual model into the MatLab version that provided so many insights into rhythmic systems. Several others, including Mr. John Domzalski , Dr. Stacey Ake, Dr. Scott Bunce, Dr. Joanne Getsy Dr. Neal Handly, and especially Dr. Joshua Jacobs, assisted with selected chapters and their comments were very useful. Dr. Bunce, Dr. Chang Chang, and Dr. Getsy also provided figures, for which I am grateful as well. I must also thank Mr. Sean Brown, who assisted with many of the figures that came from prior publications. Of course, I also must thank my long-suffering publisher, Mr. Joel Claypool, whose patience in waiting for this was remarkable. The patience of my wife, Barbara O'Donnell, and son Christopher was even more remarkable and I am grateful for the time they gave me to complete the manuscript. I would also like to thank Mr. Jay Bhatt as well for suggesting this project. For the myriad of others, including students, who commented on parts or suggested revisions, I am very grateful for your insights.

I would like to acknowledge WorldofQuotes.com (http://www.worldofquotes.com/topic/Time/5/index.html), the source of most of the quotes at the beginning of the various chapters.

Finally, and perhaps most important, I would like to express my deepest appreciation to the students, staff, and faculty of Drexel University, where I have worked for the past 28 years. They have been incredibly tolerant and understanding as I have attempted to learn how to teach. I must confess, however, that I have probably learned far more from them than I was ever able to impart in the classroom. It is a debt I will never be able to repay.

Donald McEachron
September 28, 2012

CHAPTER 1

Time and Time Again

Overview

This chapter introduces the *three basic concepts* of biological time: *Evolutionary time*; *Developmental time*; and *Rhythmic or Cycling time*. The latter is the main focus of this text and involves the evolution of *biological rhythms*. Initially, we will explore *two fundamental reasons* for the existence of these biological rhythms. The first lies in the need for any complex goal-driven device to be able to *organize and order the various activities* in which the device engages. This is not unique to biological organisms but is a more general principle applicable to any device, evolved or designed. The second reason lies in the nature of the Earth's environment and the role played by the *many geophysical cycles* in changing environmental conditions. Daily and seasonal cycles generate large environmental changes, and the ability of organisms to *adapt to these temporal cycles* will determine an organism's evolutionary success or failure. As we investigate these ideas, we will discover that the *very nature of physiological systems makes them prone to rhythmicity and oscillation*. We will also discover that the close linkages between biological systems at every level, from biochemical process to complex ecosystem, generates a *temporal cascade* in which a significant oscillation at one level of a biological system promotes oscillations at all other levels. Finally, the concept of *mathematical modeling* of biological systems is introduced as a method of checking hypotheses and assumptions.

To every thing there is a season, and a time to every purpose under the heaven:
> *A time to be born, and a time to die; a time to plant, and a time to*
> > *pluck up that which is planted;*
> *A time to kill, and a time to heal; a time to break down, and a time to*
> > *build up;*
> *A time to weep, and a time to laugh; a time to mourn, and a time to*
> > *dance;*
> *A time to cast away stones, and a time to gather stones together; a time*
> > *to embrace, and a time to refrain from embracing;*
> *A time to get, and a time to lose; a time to keep, and a time to cast away;*
> *A time to rend, and a time to sew; a time to keep silence,*
> > *and a time to speak;*

A time to love, and a time to hate; a time of war, and a time of peace.

- Ecclesiastes (ch. III, v. 1–8)

1.1   TIMING IS EVERYTHING

"All in due time"; "A stitch in time saves nine"; "It's only a matter of time"; "Time heals all wounds"; "Timing is everything." Humans live in a society dominated by time and timing. The technology of time is everywhere from the cell phone in your pocket to the computer on your desk to the clock on the wall. As I stand in my kitchen, there is a clock glowing on my microwave oven, another clock displaying a slightly different time on my stove (I can never quite get those two clocks synchronized for some reason), and yet a third digital clock hanging on the wall. This third clock is wirelessly linked to the United States Atomic clock so I know at least one clock in the kitchen is accurate (it also supplies the current date, so I *really* know what time it is). Everywhere you turn, there is yet another reminder of what time it is—often leading to a sense of panic because there is, after all, never enough time.

Ironically, all of this technology may have the paradoxical result of decreasing the average person's appreciation of time's impact on biological systems. After all, even though the indicator of a clock changes with time, the mechanism does not really seem different at different times of day. If you know how to set your alarm clock at 6pm, you do not expect to have to learn a new method of setting it if you were to try again at 6am. Despite the fact that your laptop computer has a built-in clock, it is the same computer when you are trying to answer your e-mails at 4am in the morning or 8pm at night. And therein lies the problem—the alarm clock may be the same, and the computer, but you are not. Biological organisms cycle; they change over time in frequencies as short as milliseconds to times as long as years and beyond. Thus, while the technician repairing your computer can rely on dealing with the exact same mechanism no matter when he or she works on the device, a health care worker does not have the same luxury. You are different at 6am and 6pm—the effects of a treatment or therapy at 4am in the morning can and does have radically different effects from that exact same treatment or therapy applied at 8pm at night. Human beings, like almost every other life form on Earth, display daily rhythms in physiology and behavior.

To fully appreciate this, suppose that your computer was designed with a similar daily rhythm. In the morning, a right click creates a specific drop down menu but in the evening, the very same menu is obtained with a left click. The clicking speed changes, from a slow speed in the morning to a fast speed in the afternoon to a very slow speed in the evening. Clicking on the symbol "W" initiates Word in the morning but generates Web access in the evening. It becomes even more complex when something goes wrong. In the situation as described, clicking on "W" and getting Word is normal in the morning but abnormal in the evening. When a technician attempts to diagnose and repair any problems with your computer, the approach he or she uses will change depending on the time of day when the work is being done. Restarting the computer is the correct procedure to clear a problem in the mid-afternoon, but the exact same activity seriously damages the hard drive if attempted at 3am. Just imagine how much more complex all aspects of computer use, maintenance and repair have suddenly become. And it could be even worse by adding additional cycles—hourly, weekly, monthly, and/or seasonal to the mix. This is a technical consultant's worst nightmare—if it is 2am on Wednesday in October, the problem is probably in the software configuration and I should advise the customer to reinstall the operating system, but if it is 12 noon on Monday in March, I have to tell the customer to send in the computer because the problem is probably a fried motherboard!

Thankfully, we do not live in such a world, right? Wrong! For better or worse, this cycling, oscillating world is exactly the one in which we—and all biological organisms—do live. For health care workers, that technical consultant's nightmare is their reality—humans are different at different times of day, times of the month, and times of the year. There are even biological rhythms significantly shorter than 24 hours that impact health, and well-being. The fact is that much of health care has been living in a dream world where humans—like computers—can be treated as time-invariant machines: a dream world where treatments given at different times always have the same effect and that data gathered at one time applies equally well at all times. It is a fantasy that has led to untold suffering and needless death and it is time for us to wake up.

1.2   AN INTRODUCTION TO BIOLOGICAL TIME

This book, and the subsequent volume, is intended for everyone for whom time and timing are issues of importance. This includes health care workers and engineers, truck drivers and airline pilots, military planners and civilian plant managers, professional athletes and educators—in short, just about anyone interested in the productivity, health, and well-being of human beings. In writing this volume, I am asking you to invest some of that most precious of commodities—your time. In a sense, this is a bit ironic for the focus of this book is biological time, or rather, timing in biological systems.

If you were to ask an average person what is meant by biological time, you are likely to get one of two possible responses. For some, such as professional biologists, the notion of time is linked primarily to biological evolution, that is, to *evolutionary time*. Mention evolution to someone and images of steamy swamps ruled over by dinosaurs or of heavy-set men wrapped in furs confronting raging Mastodons over an icy landscape are brought to mind. Evolutionary time seems remote and distant, occurring thousands or millions of years in the past with little in the way of relevant modern-day applications. Although these kinds of images are often promoted by the popular media, biological evolution is far more subtle and complex a process. The actual definition of biological evolution is "the change in inheritable characteristics of a biological population *over generations*." Thus, time in the evolutionary sense is a flexible concept and dependent on the length of an organism's generation. The shorter the generation time, the faster evolutionary change can take place. The failure to recognize this, and thus seriously underestimate the evolutionary potential of infective microbes, is one reason for the developing crisis involving microbial populations resistant to antibiotics and similar drug therapies.

Although evolutionary time *per se* is not the primary focus of this book, an appreciation of biological evolution is critical if the material being presented is to be properly understood. It is the concepts and processes of biological evolution that explain why the problems of a health care worker are so different from those encountered by a computer technician. The problem is not merely that human behavior and physiology are more complex than computer software and hardware, although that is currently the case. It is not just that we understand more about computers—having designed them ourselves—than we do about humans, although that too is true. The real difference is that humans are the product of biological evolution and not engineering design. As such, the success of our ancestors from the beginning of life itself depended on the ability of those ancestors to reproduce in the environments in which they found themselves. A major facet of those environments—and thus a critical factor determining reproductive success or failure—is the existence of geophysical cycles. Seasonal changes and day/night alterations of light and dark fundamentally alter the physical environment and place a crucial selection pressure on an organism's ability to anticipate and prepare for those changes. After three and a half billion years of such pressure, living organisms have biological clocks stamped into the very essence of their natures. Once you adopt the evolutionary perspective, you realize that no other result is possible. Temporal adaptation—and the rhythms that result from it—is a primal characteristic of living things.

A second, perhaps more common, response to a question about biological time might be to focus on life stage or *developmental time*. This is an especially likely response among parents who must strive to both nurture and understand their children as those children grow and develop as well as among those taking care of the elderly. That different developmental stages of life carry different risks, opportunities, benefits, and costs is obvious to all, and the importance of those stages in biological terms is clear. We may strive as children to accelerate growing older and as adults attempt to put the brakes on the very same process, but we are fully aware of both existence and the importance of life's developmental timing.

1.3   A PRELIMINARY CASE STUDY

As was the case with evolution, developmental time *per se* is not a primary focus of this book's discussion of biological time. That being said, biological rhythms interact with developmental time in some rather interesting ways. Parents of infants are fully aware of at least some of these issues as they struggle with the apparently random pattern of sleep and wakefulness displayed by their sons or daughters in the months after birth. Why does that happen and is there anything that can be done to speed up the process to get children to sleep throughout the night (so that parents can do the same)? On the other end of the scale, regular patterns of behavior seem to become less rigid and predictable as humans progress past 60 years of age. Daily rhythms become both less stable and less coherent. As a result, similar issues arise for caregivers to the elderly as experienced by parents of infants and very young children. Could there possibly be a link between these two life events? Do similar symptoms indicate similar underlying mechanisms? As a brief and very preliminary introduction to the practical nature of biological cycles, let us consider these two common observations concerning daily rhythms in the very young and the elderly.

The reason for the lack of coherence (i.e., stable patterns) in the sleep/wake cycles of the very young is thought to lie in the main coordinating clock—the so-called "master pacemaker"—of humans and other mammals. This clock is found in a region of the brain called the hypothalamus and is known as the *suprachiasmatic nuclei* or *SCN*. One hypothesis to explain the temporal behavior of infants is that the SCN of humans is not completely functional and integrated at birth. It takes time to fully develop the daily rhythm system and during this period, the temporal organization of the child is more varied and unstable than that observed in adults[14,20] leading to significantly sleep-deprived parents. On the other end of the scale, new evidence has suggested that aging leads to a loss of function in the SCN[5,11]. This loss of capacity may lead to a similar loss of rhythm coherence in the elderly as seen with infants and young children.

The data suggest an interpretation actually even more intriguing than this. In order to demonstrate the loss of function in the SCN for aging mammals, it was necessary to eliminate the light/dark (LD) cycle and examine the animals under constant conditions (more on these ideas later in this volume). When a powerful light/dark cycle is present, one with a large range of light intensity differences between light and dark, the SCN appears to function fairly normally even in aged animals. This leads to the following interesting possibility.

Humans evolved in the natural environment, constantly exposed to the extremely powerful cycle of daylight and nighttime darkness provided by the earth's rotation. Exterior light intensity ranges from less than a single lux (a measure of light intensity about $1/10^{th}$ the output of a candle measured at 1 foot from the source) at night to 80,000–100,000 lux at mid-day, depending on your exact geographic location. *Is it possible that the delay in SCN development in the young and the loss of SCN function in the elderly is due, at least in part, to the artificial environment in which both are housed*—inside buildings—where the light differences from daytime to nighttime range only over 300–1000 lux? Perhaps the human rhythm system, having evolved exposed to a very powerful external light/dark cycle, has difficulty sustaining itself when exposed to the weaker and less predictable LD cycle of interior lighting. Furthermore, this difficulty might only become apparent when the internal pacemaker itself is weak or vulnerable, due to immaturity or age-related deterioration. If so, a simple method exists—more powerful and regular LD cycles—to improve the quality of human life at both ends of the developmental scale.

You may be more than a bit confused at this point. What is a biological clock? What is a light/dark cycle and why does it matter? Although these issues will be discussed in far greater detail as we proceed, let me give you a brief idea of the concepts. A *biological clock* refers to a biological system capable of generating a rhythm on its own, without any external rhythm driving the oscillation. Typically, researchers use the term to refer to biological oscillators whose frequency is close to that of an observable geophysical cycle. The SCN, for example, is a cluster of nerve cells which oscillates with a frequency close to, but not exactly, 1 full cycle every 24 hours. Thus, any mammal with an intact and functional SCN will display nearly 24 hours rhythms in physiological and behavioral variables even when there is no geophysical cycle present, such as in constant darkness. The lack of any external environmental rhythm is called *constant conditions*. A *light/dark cycle* is any environmental rhythm in which one part of the cycle provides greater light intensity than another. An example would be a room in which the lights were on for 12 hours and off for 12 hours in a repeating cycle. This would be known as "LD 12/12." As part of adapting to their environment, organisms can adjust the frequency of their biological clocks to the appropriate geophysical cycle, a process called *entrainment*. An example of entrainment would be when a human being is exposed to the normal cycle of day and night, and the frequency of the SCN activity is entrained (synchronized) to the frequency of day/night cycle. This assumes, of course, that everything is working properly, which may not be the case in the very young and the very old.

You may still be a little uncertain about this, so let's try an analogy to try and clarify things a bit more. Consider the situation in the heart, with its far more familiar cardiac rhythms. Although much of the muscle tissue of the heart is capable of sustained oscillations, the control of the overall rhythm is normally restricted to specific central rhythm generators or *pacemakers* called nodes. It is these rhythm generators, such as the *sinoatrial node*, which enforce coherence on the overall system. If the nodes become incapable of sustaining the proper oscillations, it is possible to create an external oscillator—an electronic cardiac pacemaker—to help maintain the cardiac rhythms. What I am suggesting in the previous discussion is that a similar kind of situation may exist with daily rhythms. When the "node" for daily rhythms—the SCN—is too weak to sustain the system, it might be possible to maintain rhythmicity through the application of an external "pacemaker"—a high amplitude light/dark cycle.

While this may have clarified the situation with respect to the rhythms, a question remains. Having a child sleep through the night at a younger age clearly benefits the child's parents, but how does it benefit the child? Why should a loss of rhythm stability negatively affect the elderly, and how would preventing that loss improve their quality of life? What difference does it make if rhythms are stable and coherent versus those same rhythms being unstable and incoherent?

Again, consider the same question applied to the heart. If you were to ask that question of a cardiologist, his or her response would be precise and immediate—it is a matter of life and death. A loss of rhythmic coherence in the heart is the equivalent of death to a patient. This is because the rhythmic nature of muscle contraction in the heart is integral to the heart acting as a pump reliably circulating blood throughout the body. No rhythm, no pumping; no pumping, no blood flow; no blood flow, no life. The situation is simple, straightforward and relatively easy to understand.

The situation is not quite so obvious when it comes to daily rhythms, which is presumably why the true impact of these—and other—rhythms has been overlooked in the past. However, even this brief discussion suggests that the effects are far greater than evident at first glance. Thus, we now turn our attention to reasons behind the evolution of biological rhythms in the hope of uncovering the roles played by oscillations. Once this is understood, the door is opened for developing practical methods of effective intervention because, after all, a stitch in time saves nine or perhaps, many, many more.

1.4   WHY BIOLOGICAL RHYTHMS?

Cardiac rhythms represent the tip of an extraordinarily large iceberg. Almost every biological system or subsystem either cycles naturally or can be induced to cycle under the proper conditions. From neural activity in the brain[6] to metabolic processes, such as glycolysis[10,22] and the activity of mitochondria[1,2,7], from the release of endocrine signals[13] to rhythms in body temperature[18] and even suicide attempts[4], living systems march to a bewildering variety of different beats. The range of frequencies is extraordinary, reflecting the multitude of systems which generate rhythms and the number of potential functions to which these cycles might be put. These frequencies include the short *millisecond* neural and metabolic rhythms[6,22], *minute* long cycles in mitochondrial activity[7], pulsatile hormone secretion with frequencies of *45–120 minutes*[9,15,16,21], *daily* rhythms in almost every physiological and behavioral parameter[13,18,19] and beyond to *multi-day* sexual cycles, *annual* reproductive rhythms and even *multi-year* ecological rhythms[3,4,12,19]. Why so many biological cycles?

Three arguments supporting the evolution of biological rhythms will initially be presented, two of which rely on basic engineering principles and the third on environmental adaptation. This is not an attempt to fuse the idea of Intelligent Design, a basically religious concept, with the scientific hypotheses of evolution by natural selection. Rather, we can all recognize that evolution must deal with the same problems posed by physical reality that challenge the ingenuity of human engineers. As a result, over the course of many generations, living organisms may evolve similar solutions to those problems.

The first engineering explanation lies in the *nature of complex, goal-driven devices*. Consider for example, your automobile or other vehicle. As a responsible owner, you presumably wish to get the most mileage possible from that vehicle. One method of maintaining or improving good gas mileage is to keep the vehicle in "tune." Part of tuning the vehicle involves ensuring that the timing is correct. Timing? What timing? The timing of the processes in the engine. In order to function properly, the sequence of events within the engine must be ordered properly. If that order becomes unstable, then the engine losses efficiency. Gasoline is wasted. Even worse, poor timing can intensify the wear-and-tear on engine parts, increasing the rate of deterioration of the vehicle. Similar issues exist for other complex devices. Computers are often sold based upon their timing speed—their ability to accomplish large numbers of tasks without those tasks interfering with each other. If the computer were unable to use timing to keep the tasks properly ordered, the system would crash because commands would contradict each other and information would be randomly over-written in internal memory. The bottom line is this: timing is critical for the proper function of complex devices with multiple activities and/or parts that must act in a coherent fashion. Any loss of the proper timing wastes resources, interferes with function, and accelerates deterioration. Biological organisms are very complex goal-driven devices with multiple activities and specialized parts and therefore are as dependent on proper timing as any human-designed machine or device.

The second engineering explanation involves the *nature of control systems*. Consider a thermostat. You set a specific temperature (the *setpoint*) and if the room is too cold, the heater turns on and if it is too hot, the air conditioning comes on to cool it. Does the room stay at the temperature set on the thermostat? That is easy to test—if the room temperature stayed fixed on the setpoint, the heater and air conditioner would not be required and never have to turn back on (I wish I could figure out how to do this—it would save me a lot on heating and cooling costs!). In reality, these systems come on periodically to adjust the room temperature as it drifts away from the ideal level. If you track the room temperature, you will find that it tends to be rhythmic—as the temperature drops below ideal the heat comes on, but it takes a while before the heating is effective. Thus, the temperature will continue to drop. Once the heat becomes effective, the room temperature rises until it reaches the setpoint. However, the heating continues for a bit before completely shutting down and the temperature rises above the setpoint. Over the long term, the temperature oscillates around the setpoint, the range of the oscillation dependent on many factors, such as room size, system efficiency, insulation quality, etc.

This kind of control system is based upon the principle of negative feedback, one of many control system approaches used by organisms to regulate biochemical and physiological processes. Basic physical principles dictate that negative feedback systems are inherently oscillatory and it takes considerable investment in terms of resources and design to reduce the range and limit the oscillation. Consider how various changes in the components of a thermostatic control system would affect room temperature. For example, what is the impact of the lag time between the temperature reaching a setpoint and the system response?

We can illustrate this situation by modeling a room heating system with two lag times, 30 seconds and 30 minutes. Imagine we set the thermostat to 70°F. As a final bit of information, we assume that the room heats and cools at the equivalent rate of 1°F per minute. What is the impact of the two different lag times? Suppose we start with the temperature below 70°F and the heater coming on to warm the room. With a 30 second lag, the heater will continue to run for 30 seconds past the time the room reaches 70°F and thus the actual room temperature will reach 70.5°F before the heat turns off. The room will begin to cool (it is winter and the room is drafty) and the temperature drops at 1°F per minute. In 30 seconds, the setpoint is reached but there is a lag of another 30 seconds before the heater can respond. Thus, room temperature decreases to 69.5°F before the heater comes on. From then on the cycle repeats as long as the conditions remain the same. The rhythm which results has a period (the time it takes to repeat 1 full cycle) of 2 times the lag time or 1 minute and a range from peak to trough of 1°F. That is a very comfortable room. On the other hand, with a 30 minute lag time and following the same approach, the room's temperature would cycle between a high of 100°F and a low of 40°F with a period of 60 minutes. Figure 1.1 diagrams the two conditions. It should be noted that the time and temperature scales are different in the two graphs, indicating that the actual differences in the cycles are far greater than causal inspection might suggest. Indeed, using the scale in the lower graph, the pattern of the top graph would be impossible to discern.

In observing the two graphs, it is clear that a longer lag time leads to a greater range of values, a larger amplitude, and a longer cycle frequency. We will encounter this inverse relationship between frequency and amplitude again in the future. It is also evident from these graphs that many different

A

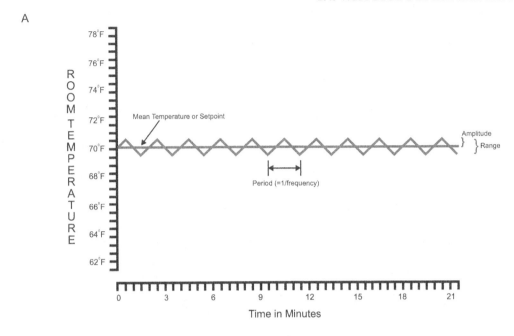

Figure 1.1: Effects of feedback and lag time on room temperature. (A) shows the effect of a 30 second lag. *(Continues.)*

frequencies can be generated from a common feedback system simply by changing the system lag time. We will encounter this situation again in future discussions as well.

For reasons that should be becoming obvious, most engineered control systems are designed to minimize unnecessary oscillations. However, there is always a cost/benefit ratio to be considered in reducing oscillatory phenomena. When the cost of reducing an oscillation exceeds the cost of the oscillation itself, there is little point to diminishing the cycle further. In many complex organizations—and all biological organisms—there are many interlocking systems, and the output of any one affects, and is affected by, numerous others. Oscillation in one such system will necessitate that all connected systems oscillate as well, leading to a time-varying structural organization. The cost of controlling all such systems to create a time-invariant system would be enormous and possibly beyond the capacity of biological evolution.

Even if evolutionary processes could generate time-invariant systems within living organisms—and almost 4 billion years of experimentation does provide for considerable creative capacity—would it be in the best interests of populations to so evolve? Up to this point, the explanations for rhythmicity and timing which were provided are common to any goal-directed, complex system whether human-created or the result of biological evolution. What makes biological rhythms different from cycles found in other complex systems is the additional factor of *adaptation to geophys-*

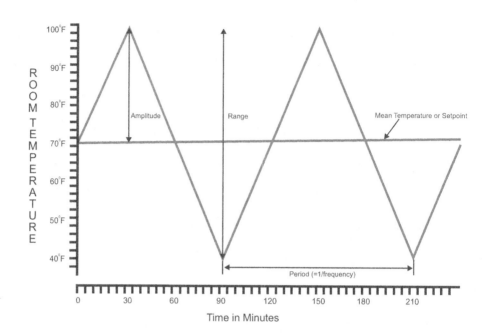

Figure 1.1: *(Continued.)* Effects of feedback and lag time on room temperature. (B) displays the results of a 30 minute lag. All other conditions are identical.

ical cycles. Even when feedback control oscillations are suppressed, all goal-directed complex systems will depend on timing—an internal temporal order. However, biological systems have evolved to adapt to geophysical cycles and thus organize their cycles into a complex pattern linked to the external environment. Thus, *biological systems must adapt to two types of temporal order—internal and external.* And the external temporal systems to which organisms must adapt are inherently oscillatory—daily cycles, lunar rhythms, seasons, and so on. In effect, one can imagine each biochemical or physiological pattern as a single instrument in a vast orchestra. Each instrument, in isolation, can provide a single musical pattern. If each instrument were played independently of what the rest of the orchestra was doing, the result would hardly be worth the price of admission. What allows a group of instruments to produce a symphony is the coordination and integration of each instrument with all the others. What ties all of an organism's biochemical and physiological rhythms together into a coherent whole are the biological clocks that have evolved to anticipate and adapt to Earth's temporal environment. It seems counter-productive to invest evolutionary capacity to suppress an *oscillatory internal* temporal order if it is critical to adapt to an *oscillatory external* temporal order. Much more likely would be the evolution of mechanisms which match internal physiological oscillations to external environmental cycles. But why should organisms evolve to match external environmental cycles?

1.4.1 TEMPORAL ADAPTATION AND THE EARLY BIRD

Perhaps an example will help illustrate this point. We all have heard the cliché "the early bird gets the worm." But is this really true—does the early bird actually get the worm? If we start by assuming a limited number of worms are available to be eaten, then it makes sense that getting to wherever the worms are before other birds in the neighborhood would be advantageous. Perhaps, arriving late would mean that all the worms have been consumed. However, this analysis is not particularly helpful since there is no time frame provided—a bird needs to be early, but early relative to what? Other birds? Ok, but then when do the other birds arrive and why? Clearly, something is missing in this analysis.

If we further speculate that worms are themselves rhythmic—that they are only available at a certain time of day, then this problem is solved. Early is relative to the time of day when the worms are available. Let's assume that the time of maximum worm availability is just after sunrise. So, the smart bird should arrive somewhere about that time. But now there is a new question: how early is early enough? Well, one possibility is to wait until there is some signal that worms are available. Then, the bird simply reacts to the signal and goes to feast on worms. But is that the optimal solution? If you wait for a signal, don't you already need to be close enough to the worms to detect that signal? Worms don't send text messages. So, how would you know when is the right time to be close enough? And if you wait until the last minute—i.e., worms are on the move—would that not give an advantage to a bird (competitor) who knew enough to be on site before the worms began to move? It would seem that waiting to react places a bird at a competitive disadvantage to a competitor who could anticipate the correct time and get there in advance.

So it would seem that the early bird does, in fact, get the worm. Except we still have not established how early is early enough. If it is advantageous to be 10 minutes early, is it better to be 20 minutes? How about 30 minutes? How about 6 hours? Is there any limit?

Clearly, there must be a limit or birds would be constantly on the ground either waiting for, or looking for, worms, and this is clearly not the case. To understand how limits might arise on this behavior, consider the costs as well as the benefits of arriving early at the worm feeding grounds. The main benefit is access to a vital food supply. But there are several costs as well. Like any other behavior, being on the feeding ground requires energy. There are also predators to consider. While the birds are after the worms, owls, foxes, cats, and others are after the birds. Also, being on the feeding ground precludes other behaviors, like mating, nest building, and protecting young. In sum, arriving early on the feeding ground requires that the bird invest metabolic energy, take the risk of being eaten by a predator, and exclude alternative behaviors in order to obtain food. The earlier the bird arrives, the longer it will have to remain and the greater the costs involved become. Thus, timing is the result of a cost/benefit analysis—too late and the benefit is reduced and too early and the costs increase. The time at which the ratio reaches its minimum point is optimal for bird arrival. Thus, it is not really the early bird that gets the worm but rather the "on time" bird. This is modeled graphically in Figure 1.2.

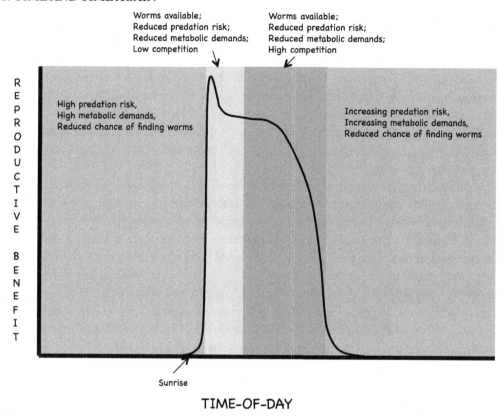

Figure 1.2: Effects of timing on the reproductive benefits of feeding on worms.

In Figure 1.2, the reproductive benefits associated with feeding on worms are plotted on the y-axis and the time of day when the feeding occurs is plotted along the x-axis. Before sunrise, the costs associated with increased risks of predation, high metabolic costs, are linked with limited benefits of feeding due to limited worm availability. Thus, there is a very low reproductive benefit to feeding on worms at that time. Just after sunrise, when the predation risk and metabolic costs go down and worms become available, there is a small spike in reproductive benefit, which is reduced as more birds arrive on site as competitors. This reduction continues as worms become harder and harder to find and, eventually, the benefits drop dramatically as predation risk and metabolic cost begin to increase again. Since the best evolutionary strategy for any organism is to maximize the number of his or her genes propagated into future generations (i.e., the "reproductive benefit" plotted along the y-axis), the best strategy for this species of bird in this environment is to time feeding for after-sunrise spike.

This example, albeit simplistic, still illustrates two important points. First, if an event is predictable, it is better to anticipate than react. Organisms which anticipate events can better prepare for those events compared to organisms which simply react to new situations as they arise. Since geophysical cycles are among the most predictable events on earth, it can be expected that biological organisms will evolve methods of anticipating tidal, lunar, daily, and seasonal events. Second, rhythmicity cascades in a system. What do I mean by that? Well, once the worms became rhythmic, then the birds which fed on them automatically became rhythmic. The predators which feed on the birds will then become rhythmic and the predators which feed on them will become rhythmic and so on. Once species within an ecosystem become rhythmic, the entire system will be forced to adopt a temporal pattern.

Rhythms do not just cascade up from species to ecosystem but down through individual organisms as well. If birds are eating worms at a specific time of day, it would be most adaptive if the bird's digestive system prepared to digest worms at that time of day. By evolving to anticipate meal timing, the digestive system could prepare the correct enzymes and transporter molecules to most effectively digest and absorb the material. The liver could adjust by pre-positioning enzymes needed for metabolism and the handling of potential toxins. In order to do this, biochemical rhythms will have to be adjusted to the daily temporal order to maximize their effectiveness. Sensory systems will change to ensure that the feeding birds are able to react to potential predators (reacting to a predator after they have eaten you is a poor survival strategy). The list of temporal adjustments goes on and on.

1.5   A CONCEPTUAL MODEL

A model may help to demonstrate how rhythms might cascade within an individual organism. Consider the representation of an endocrine chemical signaling system presented in Figure 1.3. This was adapted from a very useful text, *Problem Solving in Physiology* by Joel Michael and Allen Rovick[17]. Although very basic, it contains many of the elements that actually operate within a human body. Let's examine these essential elements.

The term "X" represents a positive stimulus, an environmental factor which causes the endocrine gland to secrete the chemical signal "H" (for hormone) into the blood. For example, X might be a cold environmental temperature and H a thyroid hormone or X could be a decrease in blood glucose and H could be cortisol. While the actual control mechanisms are more complex than this, involving multiple levels, we can model the process in this fashion for the time being and revise it later on.

Once H is secreted, it has multiple possible fates. Note first that the arrow is double-headed. This means that H can act on the very gland that secretes it. The negative sign indicates that H inhibits or decreases its own secretion. So one possible fate for an H molecule is to interact with the gland that secreted it, decreasing its own secretion. Another possible fate is to interact with the target tissue to generate some kind of biological effect(s). For thyroid hormone(s), the targets are many diffuse cells throughout the body and one of the actions is to increase metabolic activity leading to an

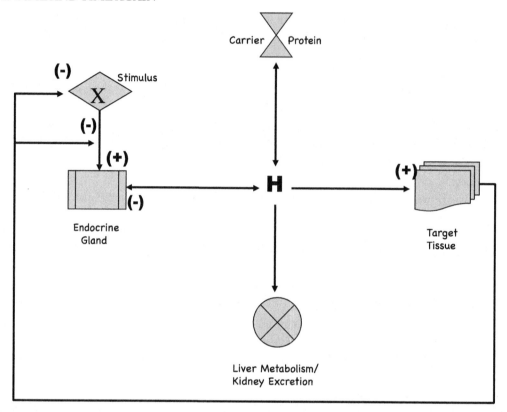

Figure 1.3: A simple endocrine model system.

increase in heat production. Cortisol acts on cells found in the liver to promote gluconeogenesis—the creation of glucose from amino acids. This leads to an increase in serum glucose levels. Note that the action of H on the target is related to the stimulus X such that the reaction of the target cells or tissue compensates for whatever effect is generated by X. If X is cold, H leads to warming; if X is a decrease in serum glucose the H leads to an increase. In the diagram, the changes in the target are linked back to the endocrine gland to allow the gland to decrease its secretion of H as the target cells/tissues' response to H compensates for X.

Two other fates are possible for H. It can be altered through the action of liver or other metabolic enzymes and then secreted into the intestines, or into the urine, by the kidney(s). This is a form of signal termination—a way of stopping the effects of H on the target by eliminating H altogether. Without this mechanism, there could easily be too much H acting on the target, creating an overcompensation for X. Finally, H can interact—be bound to—proteins found in the circulating blood plasma called carrier proteins. When H is thus bound, it still exists but cannot interact with either the target or the endocrine gland, nor can it be terminated by the liver and kidney. In effect,

H becomes stored in the blood supply and can be released from these proteins more rapidly than it might be secreted in its original form.

But how do rhythms cascade in such a system? Suppose the original stimulus, X, has a rhythm. In an environment that has the geophysical cycle of day and night, it is not very difficult to see how there would be a consistent alteration in environmental temperature, for example. If X is rhythmic, its effect on the endocrine gland will be rhythmic. Correspondingly, levels of H will be rhythmic. Once H is rhythmic, its effects on the target tissue and on the gland itself will also be rhythmic. This reinforces the original rhythm generated by X on the gland, creating an oscillating system. Given that H is rhythmic, the effects of liver and kidney activity will become rhythmic since these biochemical reactions can only act on H when H exists and the levels of H are rhythmic. Since the binding of H to carrier proteins depends in part on the concentration of H in the blood, then the levels of H bound to carrier proteins will be rhythmic because H is rhythmic. In sum, once X is rhythmic, all components of the system tend to become rhythmic as well.

Can we start somewhere else in the system? What happens if the activity of the liver enzymes in preparing H for excretion by the kidney is rhythmic? If that is true, then the levels of H will be rhythmic. As described above, once H is rhythmic, its effects on the target tissue and on the gland itself will also be rhythmic, creating an oscillating system. Again, the activity of the carrier proteins follows suit and the whole system becomes rhythmic.

Note that it is the activity of the various components that all become rhythmic, not necessarily the actual levels of the proteins and other molecular components. So, while the actual *levels* of H always cycle, this cycle only necessarily generates a similar cycle in the *activity* of liver enzymes and *binding* of H by carrier proteins—the absolute concentration of those enzymes and carriers need not change over time unless they were already cycling.

Is there any reason to suspect that levels of enzymes and other molecular components might change in a rhythmic environment even if it is not absolutely required by the model? The answer is yes. Proteins in their various forms require both energy and time to manufacture and activate. In addition, they have a limited life expectancy—proteins also degrade over time. It wastes resources, therefore, to manufacture, activate, and maintain proteins (enzymes, carrier proteins, etc.) at a constant rate if the tasks to which they would be put vary over time. It is also quite likely that the raw materials— amino acids—and energy are limited and vary themselves over time, making the temporally constant manufacture of all proteins not only inefficient but practically impossible.

The bottom line is this: no matter where in the system you start, with what component or even subcomponent, once that component or subcomponent is made to oscillate, the entire system has a tendency to oscillate in response. The interdependence of cells, tissues, individuals, and species is such that once rhythms are established at any level, they will automatically cascade throughout the entire biological community at all levels of organization.

1.6   A DYNAMIC MODEL

How do we go about the task of proving any of these assertions about the inherent tendency of biological systems to cycle? The best way would be to undertake experiments with living organisms and establish—or not—the validity of these ideas. In later chapters, I will present a vast array of experimental evidence to support these notions. However, there is another way to examine these ideas, a kind of proof-of-concept approach often used in biology when a researcher wants to bridge the gap between conceptual models and experimental evidence before committing resources to actually conduct the necessary experiments. This method converts conceptual models into mathematical models or computer simulations to determine if the conceptual models can be supported in a virtual environment. Thus, one can envision the steps of the process as follows: invent a conceptual model, convert to a mathematical form, develop a computer model, examine the results of a computer simulation in a virtual world, and then finally, experimentally determine to what extent the virtual world actually corresponds to reality.

Thanks to modern computer systems and sophisticated modeling software, it is possible to create mathematical simulations of the physiological feedback system I provided in Figure 1.2 and then ask if the model acts in a manner consistent with our predictions. If the model is created without any pre-conditions as to whether or not it will oscillate, we can at least establish the feasibility that living control systems will oscillate by determining if the model generates cycles when properly "stimulated." This approach does not provide proof of our assertions, but rather a proof-of-concept that the assertions are at least possible. If the output of the model remains stubbornly linear, then our ideas can be called into serious question. If the model output is rhythmic, then we are justified in pursuing the question further with living systems.

To create the mathematical model using MATLAB, I recruited biomedical engineering majors from Drexel University's School of Biomedical Engineering, Science, and Health Systems. In part, this was done to ensure the integrity of the modeling process, so that the models that were created reflected just the physiological systems and did not suffer from any expectation of a cycling output. Another reason is that students' abilities in this regard vastly exceed my own. When I received my doctorate in neuroscience—in the last millennium—mathematical modeling of this type was in its infancy. Thus, my desire to use such approaches is far greater than my skill, and I am forced to rely on some really bright—and far more sophisticated—students.

The model described here was designed and implemented by Mr. Kevin Freedman. The simulation was based upon the physiological control system displayed in Figure 1.4. As you can see, Kevin added a few new parameters to the previous version, such as chemical reactions transducing the effect of the stimulus, X. These enhancements increase the realism of the model and thus should not affect the proposition that such systems are inherently oscillatory.

Since much of what follows hinges on the claim that biological systems are inherently oscil-latory, that was the first idea to be examined using the model. The hypothesis is that in biological systems, as represented by the model in Figure 1.3, if any single component were to cycle or oscillate, the entire system would begin to oscillate as a result. This assertion was explicitly tested with the

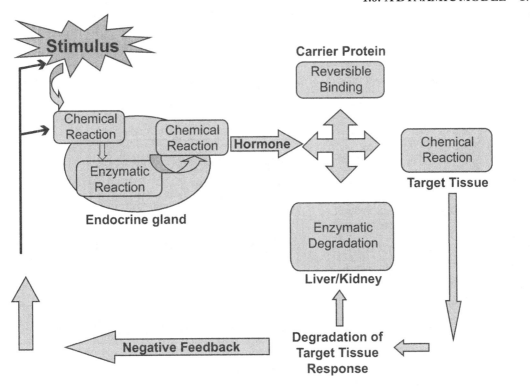

Figure 1.4: Model diagram displaying the flow of reactions taking place in order to transduce a stimulus into a target tissue response.

results displayed in Figure 1.5 Kevin plotted the output of the target tissue as a function of rhythms in various components of the system. In every case, having one component of the system cycle results in oscillations in the target tissue output. In the graph in the upper left corner, the system is initiated at time zero by being provided with a setpoint similar to setting the temperature using a thermostat. After some transient oscillations, the system stabilizes at that level. This is the situation when no component of the system is set up to oscillate. However, if any single component of the system is inherently rhythmic or is forced to become rhythmic, the whole system becomes rhythmic as a result. These results are displayed in the other graphs.

The model generated another rather interesting result when a single input stimulation from X is provided. In this particular simulation, the stimulus was applied at time zero and allowed to decay as the system responded. When the negative feedback from the target tissue was allowed to have a strong effect on the stimulus, oscillations were produced throughout the system. These results are shown in Figures 1.6–1.10.

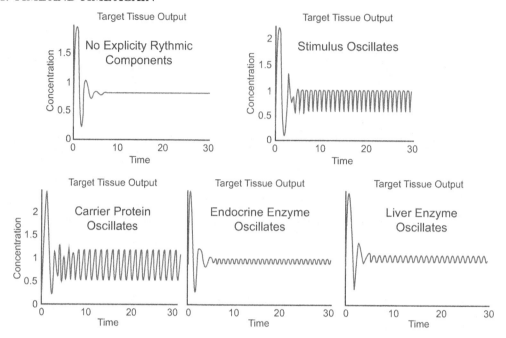

Figure 1.5: Target tissue response plots showing the effects of when a single component of a system becomes rhythmic. In this case, the target tissue output is being used to determine if the system becomes rhythmic in response to oscillations in any single component.

In this model, an environmental stimulus generates a corresponding biologic signal in order to elicit a response compensating for the original stimulus. This transduced biological signal is the stimulus of interest and is represented in Figure 1.6 as concentration (y-axis) as a function of time (x-axis). The amount of stimulus present in the bloodstream depends on the intensity of the environmental stimulus, the ability of the stimulus to reach endocrine cells and be consumed, and the amount of the end product which acts to inhibit the stimulus (negative feedback).

In Figure 1.7, the concentration (y-axis) of the cellular enzymatic activity (cellular response) is plotted above as a function of time (x-axis). Changes in enzyme activity in response to the stimulus represent the mechanism acting to stimulate the endocrine gland to secrete the hormone (H). Note for particular interest the oscillatory response in endocrine signal concentration.

Figure 1.8 shows that once the endocrine gland has been stimulated, it begins secreting the hormone into the bloodstream. The concentration of this hormone (y-axis) is plotted as a function of time (x-axis). The circulating hormone has the potential fate of being excreted, being bound to a carrier protein in the blood, being consumed by tissues, or acting to inhibit its own secretion by the endocrine gland. Again, note the oscillations in hormone concentration.

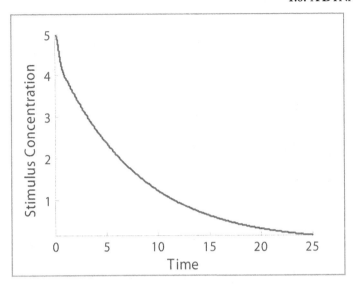

Figure 1.6: Stimulus concentration in bloodstream. A plot of the amount of biological stimulus signal over time after a single pulse of environment stimulus and the resulting transduction of that stimulus and negative feedback from the output of the target tissue.

Carrier proteins in the bloodstream act as a buffer to hold onto hormones until there is a need for them, i.e., when hormone concentration is low. When the carrier protein becomes associated with the hormone, they form a carrier protein-hormone complex (CP-H). Figure 1.9 shows the concentration of bound carrier protein/hormone and how it changes with time (x-axis). As was discovered above for other components in the system, the concentration of carrier protein bound to hormone oscillates over time.

Finally, the hormone in the bloodstream acts to stimulate a population of target cells in the body. The response to this stimulus is plotted above as concentration (y-axis) changing with time (x-axis). This output subsequently acts through negative feedback to inhibit earlier processes in the pathway. The results are shown in Figure 1.10, demonstrating that target tissue output oscillates along with the other components of the system.

The results of Kevin's model clearly demonstrate the potential of biological control systems to display oscillations. We continued to analyze the phenomena by looking at the effects of establishing setpoints on system behavior. Previously, a stimulus was provided and the system was allowed to simply react to the stimulus. In the next version of the model, the system is given a setpoint to maintain. Examples of these kinds of phenomena included physiological systems to maintain body temperature or plasma glucose levels. In Figures 1.11–1.15, Kevin provides concentration vs. time plots for key components of an endocrine gland releasing a hormone in response to a stimulus. This system in particular is responding to a stimulus which is initially zero (at T=0), and is then

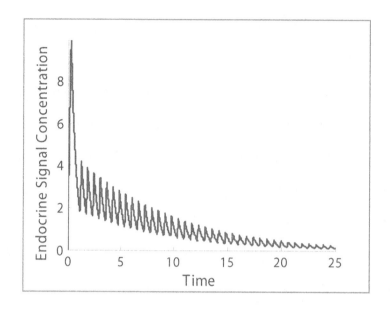

Figure 1.7: Intermediate cellular response of endocrine gland to stimulus.

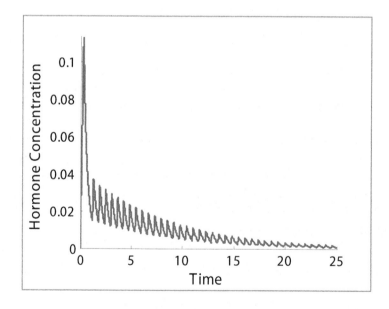

Figure 1.8: Circulating hormone concentration in bloodstream.

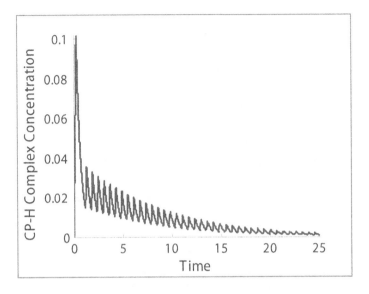

Figure 1.9: Carrier protein-hormone complex in bloodstream. The values reflect the binding of hormone with plasma carrier proteins.

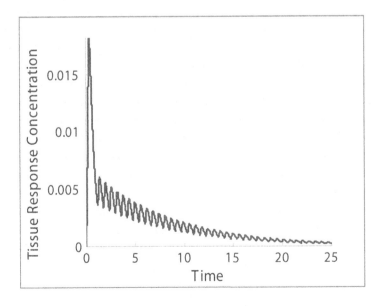

Figure 1.10: Target tissue output in response to hormone stimulation. The values display the level of response of the target tissue to hormonal stimulation over time.

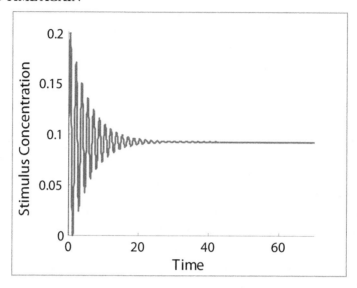

Figure 1.11: Stimulus concentration in bloodstream. This figure corresponds to the situation diagrammed in Figure 1.6 with the additional parameter that the stimulus level has a setpoint to maintain. Note the new oscillatory behavior.

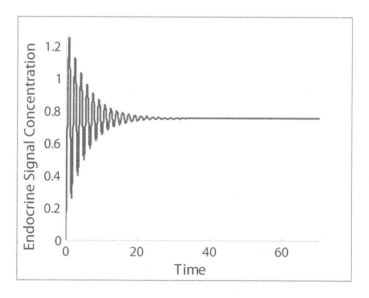

Figure 1.12: Intermediate cellular signal of the endocrine gland. The endocrine signal concentration is plotted on the y-axis as a function of time (x-axis). This figure corresponds to the situation diagrammed in Figure 1.7 with the additional parameter that the stimulus level has a setpoint to maintain.

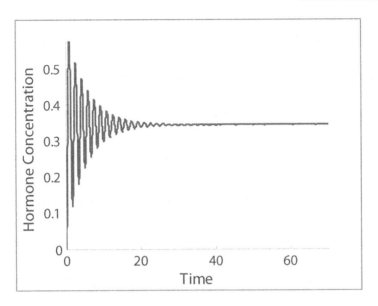

Figure 1.13: Circulating hormone concentration in bloodstream. This figure corresponds to the situation diagrammed in Figure 1.8 with the additional parameter that the stimulus level has a setpoint to maintain.

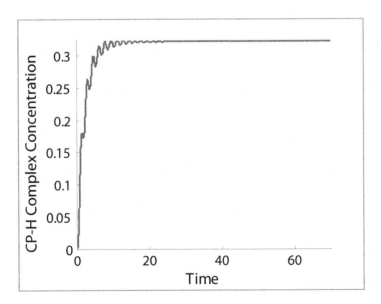

Figure 1.14: Bound hormone levels in the bloodstream. This figure corresponds to the situation diagrammed in Figure 1.9 with the additional parameter that the stimulus level has a setpoint to maintain.

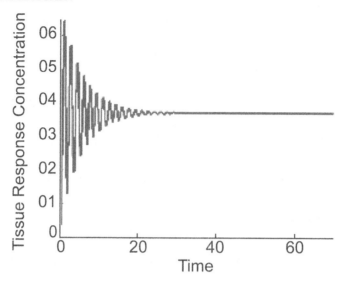

Figure 1.15: Target tissue response to hormone. This figure corresponds to the situation diagrammed in Figure 1.10 with the additional parameter that the stimulus level has a setpoint to maintain.

given an absolute set point. After a transient period of oscillations around the equilibrium value, the system becomes stable. In this new model, even the stimulus concentration shows oscillatory behavior (Figure 1.11). After an initial unsteady state where the concentration oscillates around an equilibrium value, the stimulus concentration becomes stable.

As before, the stimulus is transduced by the cellular machinery of the endocrine gland. Upon stimulation, a signaling pathway is activated which tells the cells of this gland to secrete a particular hormone. Figure 1.12 shows the effects of the stimulus on the endocrine signaling pathway.

As the stimulated endocrine gland secretes hormones directly into the bloodstream, concentration of hormones in the blood begins to rise. Secreted hormones can potentially be bound by a carrier protein, secreted, degraded, be taken up by cells, and can also act to inhibit its own secretion. Taking all these phenomena into account, Figure 1.13 shows how the concentration of hormones (y-axis) changes with time (x-axis).

Once a hormone is secreted into the bloodstream, some portion of that hormone forms a complex with carrier proteins. This carrier protein-hormone (CP-H) complex is able to hold on to or give up the hormone as needed depending on the level of hormone in the blood. The concentration of the CP-H complex in the blood (y-axis) is plotted as a function of time (x-axis) in Figure 1.14. As I indicated previously, this not a plot of the absolute concentration of a carrier protein in the blood—which may or may not be oscillating—but rather is a graph of the concentration of a carrier protein-hormone complex, which does apparently display a tendency to cycle under these conditions.

Active hormone that is not bound by carrier protein acts to stimulate a response in cells of a target tissue. The response from the target tissue is plotted in Figure 1.15 as a function of time (x-axis). This response from the target tissue acts through negative feedback to inhibit earlier processes in the pathway.

An interesting feature revealed by the model is the existence of so-called "damped oscillations" in the system as displayed in Figures 1.10–1.15. What this model suggests is that removal of an organism from a periodic stimulus will not results in an immediate cessation on internal cycles. The biological rhythms will persists for many cycles while slowly decreasing in amplitude.

1.7   RHYTHMS AND EVOLUTION

As stated previously, it is certainly possible to design and build systems that minimize these oscillatory cascades associated with feedback control. However, this is not a trivial engineering feat and requires considerable planning and detailed construction. Even so, such efforts do not always succeed. To expend the effort and resources needed to do this, the benefit must be sufficient to justify the expenditure. What about living systems? Would the benefit be sufficient to lead to the evolution of compensated, non-oscillatory systems?

To answer this question, consider again a living organism in the context of a natural environment. That environment is dominated by at least two very powerful geophysical cycles—the light/dark cycle of day and night and the annual seasonal cycles. Consider Figure 1.16, which displays the changes in temperature and humidity recorded at the Black Rock Forest, a 3,750 acre preserve on the west bank of the Hudson River, approximately 50 miles north of New York City. The data are from Dr. Kim Kastens of the Lamont-Doherty Earth Observatory and reflect temperature and humidity data collected at different elevations within the preserve. Note the substantial differences in humidity and temperature associated with the cycle of day and night. Plainly, there is more to the day/night cycle than just changes in light intensity. The entire physical environment tends to shift with the light/dark (LD) cycle. In terms of the models presented in this chapter, this is the equivalent of not just one X being rhythmic but an entire collection of various X's—of various environmental parameters—being rhythmic. Inevitably, the biochemical and physiological control systems associated with these parameters will be forced into cycling to cope with the periodic nature of the input stimuli.

Given the periodic nature of an organism's environment, creating compensated, non-oscillatory systems makes no adaptive sense. Why expend considerable resources to compensate for physiological periodicity when the periodicity of environmental stimuli makes physiological cycling inevitable? In fact, creating non-oscillatory systems is positively detrimental under such conditions. If X is rhythmic, the most successful and sustainable strategy would be to evolve rhythmic control systems which can anticipate changes and allocate resources accordingly.

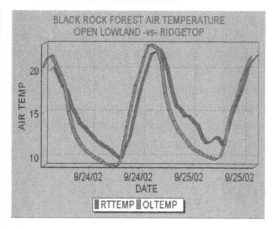
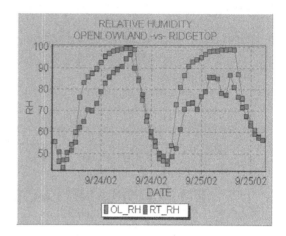

Figure 1.16: Daily changes in temperature and humidity at two elevations in Black Forest, New York. (`http://www.ldeo.columbia.edu/edu/DLESE/BRF/orientation/realtime/` created by Dr. Kim Kastens of the Lamont-Doherty Earth Observatory.)

1.8 CHAPTER REVIEW

In summary, we have approached the problem of biological rhythms from several different perspectives. These approaches can be made into a set of propositions as follows:

1. Complex, goal-directed devices must maintain an internal temporal order to function efficiently and effectively;

2. Control systems involving negative feedback are inherently oscillatory in nature;

3. Geophysical cycles impose cycles in important environmental stimuli which affect biological organisms;

4. Once an important component or subcomponent of an interconnected system becomes rhythmic, the entire system has a tendency to become rhythmic.

To these propositions, we can add one more. Many of the goals and subgoals of living things are themselves cyclic in nature. A person does not breathe once, nor move a leg once to walk a mile, nor eat one meal during a lifetime. Thus, there is a fifth proposition:

5. The nature of the interaction between living things and their environment appears to be inherently repetitious.

If we consider these five propositions as premises of a logical argument, then what is the conclusion? The only conclusion is that biological systems will be inherently rhythmic. Thus, in the case of biological systems, the cliché is absolutely true: Timing really is everything!

REFERENCES

[1] Aon, M.A., Cortassa, S. and O'Rouke, B. (2008). Mitochondrial oscillations in physiology and pathophysiology. *Advances in Experimental Biology and Medicine* 641: 98–117. DOI: 10.1007/978-0-387-09794-7_8 7

[2] Aon, M.A., Cortassa, S., Marban, E. and O'Rouke, B. 2003). Synchronized whole cell oscillations in mitochondrial metabolism triggered by a local release of reactive oxygen species in cardiac myocytes. *The Journal of Biological Chemistry* 278(45): 44735–44744. DOI: 10.1074/jbc.M302673200 7

[3] Aschoff, J. (1981a). Annual rhythms in man. In Aschoff, J. (ed.) *Handbook of Behavioral Neurobiology 4: Biological Rhythms.* Plenum Press: New York, pp. 475 – 487. 7

[4] Aschoff, J. (1981b). A survey on biological rhythms. In Aschoff, J. (ed.) *Handbook of Behavioral Neurobiology 4: Biological Rhythms.* Plenum Press: New York, pp. 3–10. 7

[5] Aujard, F., Cayetanot, F., Bentivoglio, M. and Perret, M. (2006). Age-related effects on the biological clock and its behavioral output in a primate. *Chronobiology International* 23(1&2): 451–460. DOI: 10.1080/07420520500482090 5

[6] Buzsaki, G. (2006). *Rhythms of the Brain*. Oxford University Press: Oxford, Eng. DOI: 10.1093/acprof:oso/9780195301069.001.0001 7

[7] Cortassa, S., Aon, M.A., Winslow, R.L. and O'Rouke, B. (2004). A mitochondrial oscillator dependent on reactive oxygen species. *Biophysical Journal* 87: 2060–2073. DOI: 10.1529/biophysj.104.041749 7

[8] Davis, F (1981). Ontogeny of circadian rhythms. In Aschoff, J. (ed.) *Handbook of Behavioral Neurobiology 4: Biological Rhythms.* Plenum Press: New York, pp. 257–274.

[9] Dorste, S.K., de Groote, L., Atksinson, H.C., Lightman, S.L., Reul, J.M.H.M. and Linthorst, A.C.E. (2008). Corticosterone levels in the brain show a distinct ultradian rhythm but a delayed response to forced swim stress. *Endocrinology* 149(7): 3244–3253. DOI: 10.1210/en.2008-0103 7

[10] Higgins, J. (1964). A chemical mechanism for the oscillation of glycolytic intermediates in yeast cells. *Proceedings of the National Academy of Science* 51: 989–994. DOI: 10.1073/pnas.51.6.989 7

[11] Hofmann, M.A. (2000). The human circadian clock and aging. *Chronobiology International* 17(3): 245–259, DOI: 10.1081/CBI-100101047 5

[12] Koukkari, W.L. and Sothern, R.B. (2006). *Introducing Biological Rhythms*. Springer: New York. 7

[13] Krieger, D. T. (1979). *Endocrine Rhythms*. Raven Press: New York. 7

[14] Laurinova. K and Sumova, A. (2006). Development of the mammalian circadian system. *Cesk Fysiol.* 55(4): 148–154. 5

[15] Marshall, J.C. and Griffin, M.L(1993). The role of changing pulse frequency in the regulation of ovulation. *Human Reproduction* 8 (suppl. 7): 57–61. DOI: 10.1093/humrep/8.suppl_2.57 7

[16] Meredith, J.M., Turek, F.W. and Levine, J.E. (1998). Effects of gonadotropin-releasing hormone pulse frequency modulation on the reproductive axis of photoinhibited male Siberian hamsters. *Biology of Reproduction* 59: 813–819. DOI: 10.1095/biolreprod59.4.813 7

[17] Michael, J.A. and Rovick, A.A. (1999). *Problem Solving in Physiology*. Prentice-Hall: Upper Saddle River, NJ, pp. 137–138. 13

[18] Moore-Ede, M.C., Sulzman, F.M. and Fuller, C.A. (1982). *The Clocks that Time Us*. Harvard University Press: Cambridge, MA 7

[19] Refinetti, R. (2006). *Circadian Physiology, 2nd Edition*. Taylor and Francis: Boca Raton, FL 7

[20] Seron-Ferre, M., Valenzuela, G.J. and Torres-Farfan, C. (2007). Circadian clocks during embryonic and fetal development. *Birth Defects Research (Part C)* : 81: 204–217. DOI: 10.1002/bdrc.20101 5

[21] Simon, C. and Brandenberger, G. (2002). Ultradian oscillations of insulin secretion in humans. *Diabetes* 51(Suppl. 1): S258-S261. DOI: 10.2337/diabetes.51.2007.S258 7

[22] Yang, J.-H., Yang, L., Qu, Z. and Weiss, J.N. (2008). Glycolytic oscillations in isolated rabbit ventricular myocytes. *The Journal of Biological Chemistry* 283(52): 36321–36327. DOI: 10.1074/jbc.M804794200 7

CHAPTER 2

Walking on Air: An Empirical Proof-of-Concept

Overview

This chapter investigates three propositions that can be derived from the material presented in Chapter 1. The *first proposition* is that all living organisms will display significant rhythmicity. The second and third propositions are derived from the first. The *second proposition* states that impact of an environmental factor will depend on the time at which an organism encounters that factor. The *third proposition* contends that disruption of an organism's biological rhythms will be harmful to the organism's function. In order to investigate these three propositions, the chapter focuses on a subset of all biological rhythms, daily or 24 hour cycles. In the process of investigating the evidence, we are forced to evaluate the nature and validity of various forms of *statistical testing* and how evolutionary biology provides support for using animal experimentation to investigate human behavior and physiology. We will use the problem of associating smoking and lung disease as an example of how these issues have been confronted in the past. We will also characterize the nature of *ultimate* and *proximate function* in evolutionary terms and show how *evolution permeates all questions in biology*, including those involving biological rhythms.

> *I can't blame you for finding it hard to believe, since it is in direct opposition to everything you've understood to be true in the past. It's like altering a natural law. As if I gave you proof that gravity didn't really exist, that it was a force altogether different from the immutable one we know, one you could get around when you understood how. You'd want more proof than words. Probably want to see someone walking on air.*
> *-Jason dinAlt from the novel Deathworld by Harry Harrison*[13]

At this point, you might find yourself in the same position as the audience to whom this fictional character is speaking. The arguments and simulations presented in Chapter 1 may seem correct, the logic appears sound, but you just cannot accept the conclusion. There must be something wrong, somewhere, because the conclusion just does not fit with your experience and training. Physiology and medical textbooks occasionally mention biological rhythms—and discuss some rhythms such as cardiac cycles extensively—but the overall coverage is minimal. If biological rhythms were as critical as these arguments suggest, then medical and nursing schools, biology programs, even business schools and engineering programs would all have specific courses and materials to integrate rhythms research, focusing on changes in system dynamics with respect to time, into their curricula. Treatments would be based upon optimum timing, assessing the health risks of pollutants would incorporate when the exposure took place (time-of-day, season, etc.), working schedules would be modified to promote productivity and prevent accidents, educational experiences would use timing to create the finest learning environments, and so on. Since none of these approaches are in widespread use, biological rhythms may be interesting but cannot be as important to human health and welfare as is being claimed in Chapter 1.

In the novel cited above, Jason dinAlt was able to prove his point with a single overwhelming example. Unfortunately, I do not have the advantage of Mr. Harrison's fertile imagination and so I must chip away at your doubt in stages. The first stage is to provide sufficient evidence for you to continue reading. The evidence I will present is not going to be conclusive just yet—it is more in the manner of an empirical proof-of-concept. In so doing, we will go beyond the arguments and simulations presented in Chapter 1 to evidence from the real world of living organisms. The goal is for you to accept the possibility that your experience and training might have missed a vital component to understanding human physiology and behavior. Maybe, just maybe, we really can walk on air.

To determine the best evidence to provide, let us review the conclusions from the previous chapter:

1. Biological systems have a tendency to be rhythmic;

2. Biological systems will evolve to take advantage of this rhythmicity by organizing biochemical and physiology processes into temporal cycles;

3. Biological systems will evolve further to use at least some of these temporal cycles to adapt to stable geophysical cycles.

Using these conclusions as premises, two additional ideas can be derived:

4. Biological systems will then evolve timing mechanisms to serve as organizing interfaces between internal cycles and geophysical rhythms;

5. As a result of these evolutionary pressures, the biochemistry, physiology, and behavior of organisms will be organized into predictable temporal patterns.

The primary implication of all of these ideas is that *the effect of an environmental variable or factor on an organism will depend on the time at which the organism is exposed to that specific variable*. It makes no difference whether the factor is a positive or negative, the introduction of that factor into a time-varying biological system will have results that are also time-varying. No other conclusion is possible.

A secondary implication is that disruption of the rhythmic organization of a living system is likely to have deleterious effects on that system. Temporal cycles are the result of the evolution of complex biological systems and these cycles presumably organize biochemical and physiological processes in an optimal fashion (or as optimal as evolutionary processes can achieve). Disruption of an optimal organization results in suboptimal functioning, and so disruption of the temporal structure of an organism should lead to a reduction in the function of that organism. Similar to operating a car out of tune, the disruptions of rhythms will decrease efficiency and increase wear-and-tear on living systems.

To support these conclusions, we are looking for three kinds of evidence.

1. *A demonstration that living systems display significant rhythmicity.* We can support several of the above conclusions simultaneously if we can demonstrate the existence of at least one widespread form of biological rhythm associated with a stable geophysical cycle. This is due to the fairly obvious argument that establishing a biological rhythm associated with an environmental cycle presupposes the existence of the underlying biological rhythm itself.

2. *Data showing that the impact of environmental factors varies with time.* Once a rhythm has been confirmed associated with at least one geophysical cycle, the next step is to demonstrate that these rhythms influence how the organism interacts with its environment. Even though it seems clear that no other possibility makes sense once rhythmicity itself is established, it is still important to determine the magnitude of the effects. It is one thing to argue that interaction of an organism with its environment is rhythmic—it is quite different to establish that these rhythms are significant enough to alter how human society should be organized.

3. *Evidence that disruption of a biological rhythm generates deleterious effects on the biological system.* Of course, this could be done trivially by using the cardiac cycle but this fails to establish the more general principle. A more convincing approach would be to use a rhythm associated with a geophysical cycle where the effects of disruption are not so clearly evident. If it can be shown that disruption of such a cycle—even in humans where adaptation to the geophysical cycle is no longer directly tied to survival—leads to damage within the organism, then the conclusion that temporal order is fundamental to the proper functioning of the biological system is supported.

I chose the geophysical cycle of day and night with its near 24-hour cycle to provide evidence for the three propositions described above. The underlying biological rhythms would be those with periods that match the 24-hour light/dark cycle. This system was chosen in part due to the critical importance and stability of the day/night cycle and in part due to the decades of work done with these rhythms. The latter provides significant data with which to test our conclusions. Before turning the evidence itself, however, I would like to review the nature of scientific data and argument to ensure that the reader can critically review the materials to be presented.

2.1   ON THE NATURE OF SCIENTIFIC EVIDENCE

It is not in the nature of scientific investigations to deal in absolute certainty. Science is based upon empirical data about which there is always some measure of doubt. In addition, the logical arguments used in scientific investigations are limited in their ability to produce total confidence, so there is always a small "grain of salt' which must be tucked away, ready to be applied to any conclusion generated through scientific methodology. A complete discussion of these kinds of issues is beyond the scope of the present volume and the interested reader is referred to other treatments, such as found in Hempel[16], Lyttleton[23], Medawar[25], and Root-Bernstein and McEachron[33].

Despite the fact that all scientific investigations carry some measure of uncertainty, the level of doubt or confidence in any conclusion varies greatly depending on the nature of the evidence. The claims of actors on such TV programs as *CSI* or *NCIS* notwithstanding, evidence can lie, or at the very least, be misinterpreted. The likelihood of such misinterpretations depends on the character of the evidence and processes by which it is gathered. To judge the quality of the evidence we will consider, it is necessary to understand at least a bit about how data can be gathered in support of scientific investigations.

We can define three categories of evidence: *anecdotal, correlational,* and *experimental.* Anecdotal evidence is based upon personal observations generated through the normal life experiences of the observer(s) and not based upon any planned investigation. Humans are very good at pattern recognition and thus are able to discern unusual events that do not fit the accepted model. The problem is that humans are so good at creating patterns that they will create models whether those models have any basis in reality or not. Thus, humans can see exceptions to a natural rule when no such rule actually exists. This means that anecdotal evidence, by itself, provides no method by which the real can be differentiated from the merely apparent. Thus, the only function of such data is to provide a justification for further research.

Correlational evidence is created when one factor or set of factors is mathematically related to another. Although far better than anecdotal evidence, such analyses suffer from an inherent weakness: a correlation between two factors cannot be used to establish a causal relationship between those factors. A classic example was the correlation created between soft drink consumption and the rates of people acquiring polio in the early 20th century. If one were to assume by such a correlation that consuming soft drinks caused polio, than all soft drinks should be banned. However, this makes no sense—how could soft drinks generate an infectious disease? There is no model to support such a notion. Why, then, does such an association exist in the data? The answer lies in the nature of correlational data. When two factors—in this case, soft drink consumption and polio—can be correlated to a third—hot summer weather—they will correlate to each other as well. The actual causal relationship was that hot summer weather led to crowding in public areas, such as swimming pools, where polio could spread. Such weather also generates an increase in the consumption of fluids, including soft drinks, leading to the correlation of soft drink consumption and the incidence of polio[10].

Correlational evidence can be exceedingly misleading and thus dangerous. It is like driving a car. A knowledgeable driver, experienced with both the vehicle and the road, has a fair chance of getting to the correct destination. A novice without any prior instruction and ignorant of both the car and the terrain, has an equally high chance of causing an accident. Accidents with correlational data are common and such evidence must be viewed with the appropriate caution before drawing any conclusions.

The final type of evidence is experimental. This is the only kind of approach where causality can reasonably be established. Unfortunately, this does not mean that all experiments give accurate and reliable results, even if done correctly. It is the nature of statistical analysis that a researcher can design and execute his or her experiment properly, analyze it appropriately, and derive well-supported conclusions and still be wrong. The probability of such errors decreases as the number of experiments generating the same results increases, but the probability never quite drops to zero. However, as the number of repetitions increases, our confidence in the results gets stronger.

There is an additional caveat to consider with biomedical research of the kind we will be discussing in this text. When designing and executing an experiment, a researcher selects a number of organisms to examine within the experimental protocol. This is a *sample*, a subset of the entire population about which the researcher wishes to discover some attribute or characteristic. If the sample is chosen properly, the results of the experiment can reasonably be applied to the whole population. The catch is the definition of a "population." Human experimentation carries risks and so many experiments are conducted, as least initially, with non-human animals. Technically, the population about which a researcher can make any conclusion is the population of the species from which he or she collected their sample. Thus, experiments on mice might tell us something about the population of all mice. This can be made even more problematic when one considers that most such experiments use laboratory bred animals as distinct from their relatives found in natural conditions. These groups can be so different that conclusions from an experiment can only be statistically justified as applying to the population of laboratory mice of the same strain!

I can remember how this problem created difficulties for a researcher discussing the results of his study with a television reporter some years ago. The researcher was reporting findings on obesity in mice and was trying, with increasing levels of desperation, to point out that the findings needed to be confirmed and extended before being applied to humans. The reporter was having none of it, however, and was cheerfully describing how the data would end human obesity in our time. Later, it turned out that this particular characteristic was, in fact, specific to the mice being studied and was not directly applicable to humans after all.

Presuming that the goal of biomedical research is not merely to improve the well-being and productivity of specific strains of laboratory mice, how can we proceed? The most obvious method is to conduct human experimentation so that the population matches the sample. However, this is not always possible. The history of research into smoking and lung disease provides an alternative approach to direct human experimentation. It also illustrates how anecdotal, correlational, and experimental evidence play various roles in studying biology and health.

In a recent review of the statistical approaches used to link cigarette smoking with lung disease, Parascandola[29] documents many of the arguments that were made against the use of correlational methods to attempt to prove a causal relationship between the two factors. Initial correlational studies examined numerous possible factors which appeared on the basis of observation (anecdotal evidence) to be potential agents generating the rise in the incidence of lung cancer in the 20th century. Follow up studies began to focus more exclusively on smoking. Eventually, the sheer number of studies, both retrospective and prospective, which demonstrated a relationship made it unlikely that any common unknown factor or selection bias could explain all the results. Even so, there was always a remote possibility of such a factor, and researchers looked to animal studies to confirm a causal link between smoking and lung disorders, including cancer. The underlying logic of the approach was that if smoking could induce lung disease in animals under controlled laboratory conditions, then the mechanisms could be ascertained and a model developed showing how smoking damaged lung function. Once such a model existed, it could then be applied to humans to determine if the changes observed in the laboratory animals matched those seen in human patients. If the changes observed in animals under controlled conditions were duplicated in human smokers, then it is a reasonable assumption that the same processes are at work in both species. Although not as strong as a controlled experiment using human subjects statistically, it is ethically the only possible method by which the problem can be resolved. Although it took far more time than most researchers expected, animal studies demonstrating a causal relationship between smoking inhalation and lung cancer and other disorders have begun to appear[18,39]. (See Hecht[15] for review of studies.)

When evaluating the evidence concerning the impact of biological cycles on human health and productivity, it is therefore vital to determine if the data under consideration are anecdotal, correlational, or experimental.

If the evidence is anecdotal, it may only serve as a starting point for an investigation—no final conclusions can be drawn from such material.

If the evidence is correlational, then we should ask two further questions:

a) Are there experimental studies which support the associations demonstrated by the correlations?

b) Is there a model or mechanism that can be shown to causally link the correlated factors?

Finally, if the evidence is experimental, then we need to ascertain the population from which the sample was taken. If that population is non-human, we will need to provide a justification for extrapolating the results from the experimental subjects to humans.

2.2   THE EVIDENCE

Proposition 2.1   *Living systems will display significant rhythmicity.*

For this proposition to be confirmed in the case of 24-hour rhythms, studies would have to be conducted demonstrating that a large number of species display such rhythms. In 1981, a number of reviews concerning biological rhythms were published in the *Handbook of Behavioral Neurobiology, Volume 4*. These papers covered the search for biological rhythms in a large number of animals, both vertebrate and invertebrate. Simply considering invertebrate behavioral rhythms, Brady[3] was able to list studies reporting daily rhythms in species ranging from single-celled Euglena to multicellular but simple sea anemones and other similar creatures (currently in Cnidaria) to nematodes, flatworms, and round worms and on to mollusks (various snails and Aplysia), echinoderms, crustaceans, arachnids, and insects. All in all, 24 different species of invertebrates were reported as displaying 24-hour cycles in various behavioral parameters from phototaxis to walking and flying. In a similar manner, Rusak[35] investigated the prevalence of daily behavioral rhythms in vertebrate species through a literature review. He was able to list studies confirming daily rhythms in 58 species of fish, 20 species of amphibians, 18 species of reptiles, 50 species of birds, and 46 mammalian species, including humans. Neither author claimed to have covered the entire literature at the time, and more species have been added to this list since 1981. In fact, the volume of observations supporting the existence of daily rhythms in animals have led some to claim that the property is ubiquitous in the animal kingdom[38 in 5].

Daily rhythms, however, are not confined to animal species. Microorganisms such as *Neurospora* and *Gonyanulax* display daily cycles (see Lakin-Thomas and Brody[21] for a recent review) as do those plant species which have been studied[24,37]. Even types of bacteria have been discovered to display 24-hour rhythmicity[11,20]. In fact, daily rhythms have been found in all eukaryotic species (organisms whose cells are organized around a nucleus containing genetic material) so far examined[21,26,38]. While there may be some exceptions yet to be discovered (obviously not all species on the planet have been studied), the general rule appears to be that complex biological organisms display daily (24-hour) rhythmicity.

There are two important consequences to these data. First, it appears that we have successfully established the validity of Proposition 2.1. Second, there are evolutionary consequences to the apparent ubiquity of these daily rhythms which has significant impact on how such systems are studied and the conclusions which can be drawn from those studies. As described above, actual experimentation is the most powerful method in the scientist's arsenal for testing hypotheses. The results of such experiments are statistically valid only when applied to the population from which the sample used in the experiment was drawn. Thus, establishing daily rhythmicity in a species of laboratory mice technically only applies to that species and cannot be automatically extrapolated to other organisms. This is true statistically and limits the utility of animal experimentation unless an argument can be made to justify an extrapolation. Here is where evolution comes to rescue the experimenter from an apparent dead end.

2.2.1   ON THE NATURE OF EVOLUTIONARY PROCESSES

Evolution is based upon the actions of mutation and natural selection. Mutation is any change in the DNA sequence of a gene and is thought to occur randomly. Random in this case is relative to the needs of the organism. For example, suppose a mutation in gene A would make an individual of a certain species more attractive to members of the opposite sex. Clearly, this would be advantageous to that individual. However, mutations are random relative to the needs of the organism and thus this mutation may or may not occur. If it does, then natural selection would favor the reproduction of that individual. Thus, while mutations may occur randomly, the effect of those mutations is quite clearly non-random. In other words, natural selection can be thought of as a kind of filter which selectively promotes the success of some genes and the failure of others.

Even the mutation process is not quite as random as might be supposed. The mutation described above was in a gene A. This presupposes that gene A already exists. Otherwise, there is no DNA sequence for A which could be changed. Since the very existence of gene A is dependent on the mutations and natural selection previously applied to the ancestors of the individual currently with gene A, the types of possible mutations in this individual are limited by those evolutionary processes which affected earlier generations.

All of this leads to the concept of descent with modification—new species evolve from pre-existing forms through the dual actions of mutation and natural selection. This can be diagrammed as shown in Figure 2.1.

But wait, what is going on here? We were discussing the validity of experiments and seem to have wandered off into evolutionary biology. What can this discussion have to do with the observation that all eukaryotic species and even some prokaryotes (such as bacteria) display 24-hour rhythms? The argument goes as follows. Suppose you conduct an experiment using species F. Strictly speaking, your conclusions will be valid only for species F. However, assume that the characteristic you were examining—say lung function—is also seen in species G. It is reasonable to suggest that since species F and G share a recent common ancestor, their common ancestor also shared this characteristic. In that case, although you would not be able to speak with strict statistical validity, you could plausibly argue that results in species F probably apply to species G as well since both inherited the characteristic from the common ancestor. Thus, evolutionary biology provides a logical basis for extrapolating experimental results from one species to others which are evolutionarily, and thus genetically, related.

Just how far can this logic be pushed? For example, suppose species F through N share a common feature. Operating from an evolutionary perspective, the conclusion would be that the common ancestor of F through N had some original form of that feature. Thus, extrapolating experimental results from F alone to F through N seems valid. If species A through E also share the feature, then the common ancestor can be pushed back to the most common ancestor in the diagram. Does that mean we can extrapolate results from F to A through N?

Unfortunately, we have no way to definitively answer this question. On the positive side, the more species that share a characteristic, the earlier that characteristic is thought to have evolved.

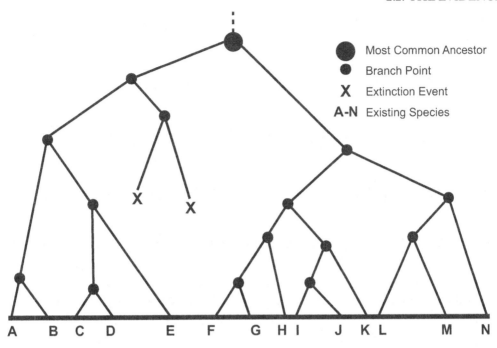

Figure 2.1: Diagram of a hypothetical evolutionary tree.

Characteristics whose similarities are due to descent from a common ancestor are called *homologous*. There are certain *universal homologies*, such as the use of a triplet nucleotide sequence to code for amino acids, which are considered to be common features of all living things[1]. In addition, the evolutionary process is inherently conservative. Once a particular process has evolved and works at a reasonable level of effectiveness, the process tends to be retained even if it might not be the ideally optimal solution. Because new genetic material comes primarily from mutation, existing processes are more likely to be modified through evolutionary means rather than reinvented from scratch. On the negative side, the farther the common ancestor for a feature is pushed back in time, the more time is available for modification. After all, the term is descent *with modification* and not just descent alone. The more time available for modification, the greater the chances that modifications will, in fact, occur.

Note a slight caveat to the above arguments. Species can evolve very similar observable characteristics without being closely related evolutionarily if the selection pressure exerted upon them is similar, sufficiently strong, and consistent. The similar body shapes of sharks and cetaceans are not thought to be due to descent from a common ancestor but rather they are the likely consequence of the species being subjected to the same physical requirements exerted by the need to swim in ocean water. These common observable characteristics when not due to common descent, are called *analogous* features.

What does all this imply for daily rhythms? If all eukaryotes truly display daily rhythms, there are only two possible explanations. Either the earliest common ancestor of all eukaryotic species developed daily rhythms as adaptation to day/night cycles and passed this adaptation on to all their descendants or the selection pressure exerted by day/night cycles caused the independent development at some point in every lineage leading to all currently surviving eukaryotic species. The latter hypothesis, while possible, seems unlikely, and molecular studies would seem to support the notion of a common ancestor[4,9,17,40]. *If true, this would make daily rhythms a universal homology for eukaryotes.* On the other hand, differences in the molecular mechanisms underlying daily rhythms in cyanobacteria indicate the possibility of a separate evolutionary event[21].

2.2.2   OBSERVATIONS ON HUMAN SUBJECTS

All of this discussion may seem be a bit nebulous. Granted, all eukaryotes have daily rhythms and humans are eukaryotes, so humans must have daily cycles. But, more specifically, what kind of daily cycles do humans display? And, more importantly, are these cycles really critical to human health, well-being, and productivity? Is the observed variation significant enough to warrant the effort needed to redesign health care delivery systems or readjust work schedules to take these cycles into account? Are daily cycles a vestigial leftover from our evolutionary history that can be safely ignored or are they are such a fundamental part of human physiology that failure to plan using daily rhythms generates needless error, pain, and even death?

To partially answer these questions, we need to examine data obtained from observing human subjects. Consider such data as are presented in Figure 2.2. These data were taken from two groups of human subjects, both healthy subjects (black lines) and patients suffering from rheumatoid arthritis (red lines). The graphs depict variation from the 24-hour mean in percentages. In other words, for each graph, the mean level of the variables was determined over a 24-hour period and then each point was graphed as a percent difference from that mean. For the moment, let us restrict our attention to the cortisol data from healthy subjects (Graph A, black lines).

According to the graph, serum levels of cortisol vary from about 40% of the mean to well over 170% of the mean. To provide some sense of the magnitude of this, suppose your body weight were to vary on the same percent scale. If your average weight were 150 pounds, at some points during the day, you would weight as little as 60 pounds while at other times, you would weigh as much as 255 pounds!

Let us pursue this analogy a bit more. Suppose weight really did change daily on such a scale. Suppose further that this change was not recognized by health care providers when weighing patients. What would happen? It seems unlikely that the variation could be overlooked entirely—it would be rather hard to miss that your patients' weights varied over a range of 195 pounds! However, what could be missed is that this change was tied to time of day, that patients weighed significantly more at certain times of day and less at others (even this seems like a stretch, but bear with me

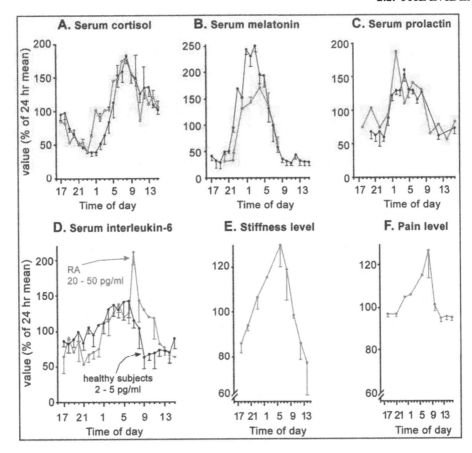

Figure 2.2: Daily rhythms in various physiological parameters in healthy subjects (black lines) and patients suffering from rheumatoid arthritis (red lines). (From Cutolo and Straub, 2007 with permission[6].)

on this). As a result, health care providers would have to set a normal range of 100–250 pounds for body weight since healthy patients clearly varied by this amount. This range is so great as to obscure any pathological weight changes until the pathology is very advanced. In engineering terms, the noise generated by enlarging the range of normal values to encompass the daily variation is so large as to drown out any signal generated by pathology until the signal becomes to great enough to extend beyond the range. The total range of expected cortisol in humans, for example, has been reported by the Medical Council of Canada in 2008 as 160–810 mmol/L reflecting the enormous daily variation in this endocrine chemical signal (http://www.mcc.ca/Objectives_Online/objectives.pl?lang=english&loc=values).

The changes observed in serum cortisol levels (and prolactin as well, see Graph C) are sufficient to require clinical laboratories to incorporate some level of temporal analysis in their reporting procedures. When collecting blood samples for serum cortisol levels, for example, it is typical to report morning (a.m.) or evening (p.m.) sampling. How much does this help? Is this sufficient to account for the rhythmic variations seen in the figures?

To answer this question, let's create a hypothetical physiological parameter with a daily rhythm as seen in Figure 2.3. The thin vertical lines indicate the expected variation in the levels of this parameter at each time in the cycle. In an attempt to compensate for the parameter's daily rhythm, samples are classified as being either from the morning (labeled AM on the graph) or evening (PM). Does this compensate for the cycle? It does certainly seem to represent an improvement. Without such a classification, the normal range would have to encompass the entire 24-hour cycle plus the variances at each point. However, there seems room for additional improvement. A sample classified as AM, with the entire morning section of the curve contributing to the variance, would have a normal range lying between the two triple red lines. In that case, the blue samples would be classified as "normal" even though they are clearly beyond the standard range of values. In fact, only the green sample would be categorized as "abnormal" in this circumstance. Without the AM-PM classification system, even the green sample would be considered normal.

Here is the bottom line. If a physiological or biochemical parameter varies over time in a predictable way, the contribution of the cycle to the overall variability of the parameters can be compensated for by taking the time of sampling into account. This reduces the noise level of the signal and increases the chance that abnormal changes will be detected. Since early detection is a key factor in the successful treatment of disease, it follows that time-of-sampling is a key metric in collecting clinical data.

What is not clear from this discussion is the exact method by which this conclusion can be applied to clinical situations. For example, with what temporal resolution do samples need to be classified? If two time points are not sufficient, how many time points are needed? This turns out to be a complex issue involving numerous parameters such as the range and amplitude of the specific rhythm, the variation at each time point, the variability of the rhythm itself and the population variances.

There are, however, additional implications from Figure 2.2 that can be discussed now. For example, the graphs labeled E and F refer to symptoms experienced by the patients suffering from rheumatoid arthritis. These symptoms are clearly rhythmic and this might seem just a bit strange. It is one thing to argue that various physiological parameters evolved daily rhythmicity to adjust to the day/night cycle, but why would the effects of diseases be rhythmic as well? What is the evolutionary advantage of cycling symptoms?

The answer may likely lie again in the conclusions about rhythms that were deduced in Chapter 1. When a complex system is linked together through dependencies and feedback, the rhythmicity of any major component of the system will tend to cascade throughout the organization and make the entire system cycle. Thus, if the physiological components affected by a disease are

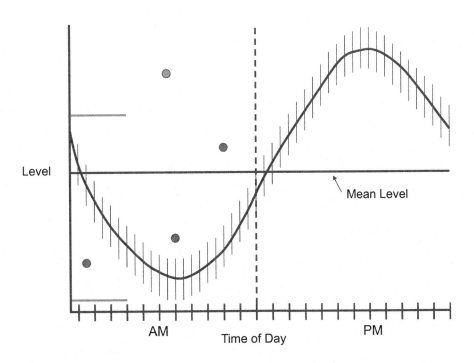

Figure 2.3: Hypothetical rhythm in clinically important physiological parameter.

themselves rhythmic, even if the disease itself is not, the effects of the disease—in other words, the symptoms associated with the disorder— will necessarily cycle. Given that many clinical treatments are, in fact, symptom relievers, predictable temporal variation in symptoms opens up the possibility of using timed treatment to reduce those symptoms.

Another example of daily rhythms in symptoms is presented in Figure 2.4. In this instance, symptoms associated with rhinitis (nasal inflammation) are examined across 24 hours. The data are categorized by gender on the left and by smoker vs. non-smoker on the right. Clearly, symptoms vary in intensity over the 24-hour sampling period. In Figure 2.5, Michael Smolensky and his colleagues[36] demonstrate that timed treatment of time-varying symptoms has the potential to create more efficacious treatment regimens.

2.3   SUMMARY AND CONCLUSION FOR PROPOSITION 2.1

Proposition 2.1 —that living systems display significant rhythmicity—has been established by demonstrating that daily cycles exist in all eukaryote species thus far examined and even for some

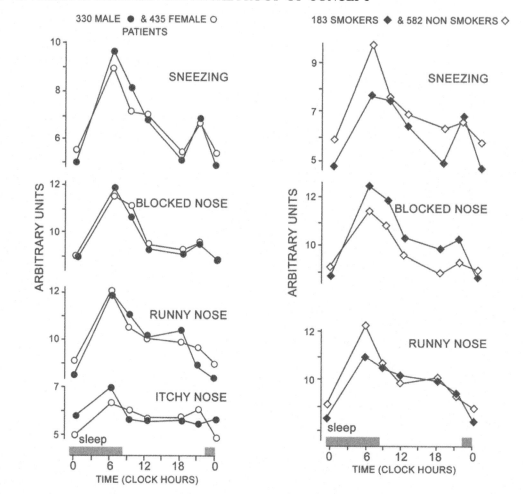

Figure 2.4: Twenty-four hour patterns of symptoms associated with rhinitis. (From Smolensky, et al., 1995 with permission [36].)

prokaryotes. Molecular evidence to date supports the hypothesis that these rhythms are homologous in eukaryotes and that the mechanisms used to generate such rhythms evolved from a common ancestor in the far distant past. Given the time involved from this putative ancestor to modern species, considerable modification of the actual molecular mechanisms generating daily cycles can be expected.

The evolutionary commonality of daily cycles allows us to use animal research to now investigate Proposition 2.2: *The impact of environmental factors will vary with time at which the organism is exposed to those factors*. This is quite useful because the kinds of experiments involved are not always

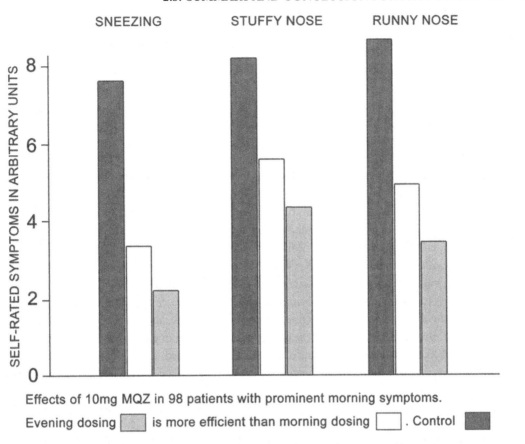

Figure 2.5: Demonstration of the efficacy of timed treatment to control symptoms that vary by time of day. (From Smolensky, et al., 1995 with permission[36].)

appropriate for use in human beings. For example, it is a common practice in the development of new pharmaceutical agents to expose animals to toxic levels of the agent in an effort to determine at what level of drug side effects will occur. On some occasions, this goes so far as to calculate something known as the lethal dose 50 or LD_{50}, which is the dose of agent at which 50% of the treated population dies. This is obviously not an experimental paradigm useable for studies with human subjects. The fundamental assumption in any of these kinds of experiments is that there is an evolutionary commonality underlying the reaction of the animal test subjects and the human patients to the drug and that this commonality allows some level of extrapolation from the animal results to humans. Although no one—except the occasional journalist—expects the extrapolation to be absolute, given the evolutionary distance and modifications associated with that distance, the insights gained from animal experiments can provide a starting point for predicting human reactions. Now that the evolutionary commonality of daily rhythms has been established, we have a firm basis for extrapolating from animal studies of daily cycles. We can confidently predict that human physiological and behavioral reactions, although not identical, should be similar enough to those observed through such animal experimentation to justify further research.

Proposition 2.2   *The impact of environmental factors on biological systems varies with time.*

We have already uncovered some evidence for Proposition 2.2 from the data displayed in Figure 2.5. Sometimes, however, it really is necessary—and possible—to have a clear demonstration and indeed to "walk on air" in order to fully appreciate the validity of an argument or proposition. This kind of demonstration is admirably shown by the data displayed in Figures 2.6 and 2.7 from the work of Erhard Haus, Franz Halberg, and their colleagues[14] published in 1974.

The data displayed in Figure 2.6 shows the results in terms of mortality and survivorship of irradiating mice with 550 rads of X-radiation at different times of day. The left graph shows the percent mortality 8 days after exposure while the right graph show mean survivorship. The results are startling, to say the least. By exposing the same strain of mice to the same environmental variable in exactly the same manner but at different times of day, the 8-day mortality varies from 0% to 100%. To be clear, the exact same radiation exposure that killed not one mouse 8 days after exposure when given at 1600 hours was 100% lethal in the same time period when given at midnight. While not quite so overwhelming, the mean survivorship also displays a considerable variation simply related to the time of exposure.

Is this some strange phenomenon unique to the effect of X-rays? Apparently not. In Figure 2.7, Haus and his colleagues show the results of exposure mice populations to the endotoxin (poison) produced by *E. coli* bacteria. Again, the data are startling. A dose which was lethal to the great majority of animals when given at one time affected only a small minority of the population when injected at another. Again, the only difference generating these results is time of day. The strain of animal, the environment, the dose, the agent, the route of administration is identical from one time to the next. What is particularly impressive about the data displayed in Figure 2.7 is that it comes from two separate experiments[12].

The next question to consider is whether or not these effects only appear when dealing with exposure to lethal or near-lethal concentrations of an environmental factor. Similarly, the clear answer is no. In Figure 2.8, the results obtained from sublethal injections of the antibiotic gentamicin are plotted as a function of time-of-day of the injection. Two hundred milligrams per kilogram of the antibiotic were injected intramuscularly into groups of rats at different times of day. The increases in excretion of three urinary enzymes were measured as an indication of renal tubule toxicity. The values are plotted as the percent increases in enzyme concentrations over a 24-hour period three days post-injection compared to similar 24-hour period pre-injection. Thus, each animal served as its own control. Statistically significant differences were found related to time of injection for all three enzymes[30].

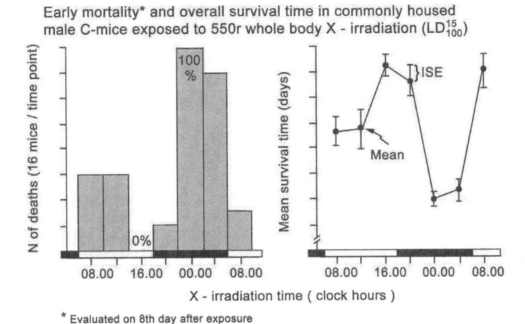

Figure 2.6: Reaction of mice to 550 r exposure of X-rays at different times of day. Left graph shows mortality 8 days after exposure while the right graph displays mean survival time of the same animals. (From Haus, et al., 1974 with permission[14].)

Finally, the last question to ask is whether or not these data apply only to laboratory animals or do they have a broader application. Despite the logic of our evolutionary arguments, nothing is more satisfying to the scientific mind as experimental confirmation. In the case of aminoglycosides antibiotics, such as gentamicin, experimental confirmation already exists.

Beauchamp and Labrecque[2] report on a number of studies involving both patients and healthy volunteers that confirm time-of-administration as a significant factor in the pharmacokinetics of many aminoglycosides. Interestingly, the highest levels of renal toxicity occur when the drugs are administered during the resting phase of the daily cycle in both animals and humans. This can be seen in Figure 2.9, which displays data obtained from both laboratory animals (the upper graph) and humans (the lower graph). The laboratory animals—rats and mice—are nocturnal and most active at night. In contrast, humans are diurnal or a day-active species. Thus, in both cases, the maximum renal toxicity is appears in the non-active phase of the cycle.

Given this evidence, the main question is why there should be such powerful daily rhythms in effects of environmental variables? Of course, we predicted this—indeed, this is the evidence we expected to find, although perhaps it is a bit more startling than at first anticipated (anyone feel

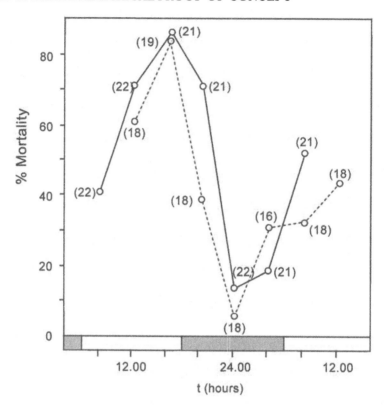

Figure 2.7: Susceptibility rhythm of mice given intraperitoneal injections of *E. coli* endotoxin at different times of day. Results are from two separate experiments. (From Haus, et al., 1974 with permission[14].)

like walking on air now?). However, the argument that the effects of environmental variables are rhythmic because the entire biochemistry and physiology of organisms is inherently rhythmic and thus, the effects of these stimuli must be time-varying as a result, is a bit unsatisfying. After all, this kind of naturalistic explanation involves identifying ultimate causality or function. Useful as that may be, it is not enough. We also want to understand the proximate mechanisms that lead to the observed results.

A brief discussion of the differences between the two levels of explanation might help clarify this. If I were to ask you what the purpose of a for-profit corporation was, without providing any other information, the only response you could give would be to reply "to make a profit." You could reply with this answer because making a profit is the universal function of all for-profit corporations—it is the *ultimate* function of all such enterprises. In a similar way, you could say that the ultimate function of all living organisms from an evolutionary point of view is to preserve and propagate genetic material. Every single living thing has exactly the same ultimate function or purpose. This

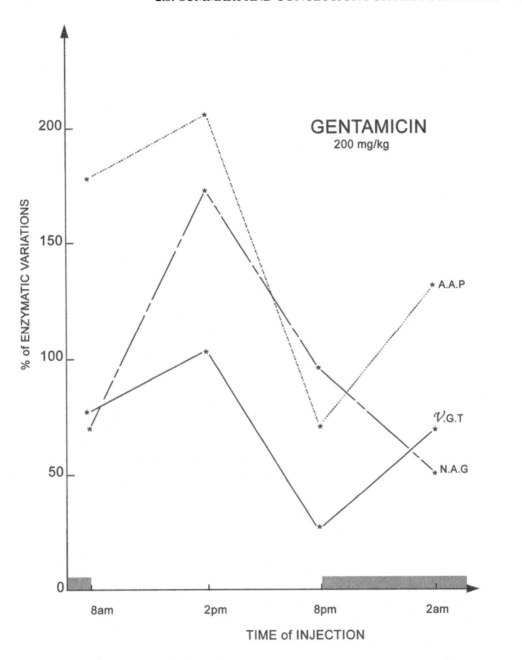

Figure 2.8: Percent increases in urinary enzymes after injection of gentamicin at different times of day compared with control values. (From Pariat, et al., 1998 with permission[30].)

is surely a vital and extremely powerful insight. However, when my garden is being overwhelmed by weeds, how does knowing that the ultimate function of weeds is to preserve and propagate weed genetic material going to help me? Actually, it seems rather obvious that weeds are fulfilling their ultimate function with great success, but I would like to actively interfere with that process, and not just sagely observe it. Similarly, if I'm trying to obtain employment at a for-profit business, it would be useful to know just exactly what the business did to obtain that profit. This is not always the case—if you are a tax accountant, simply knowing that the entity is for-profit might be enough. However, for most positions, it is vital to know how a business operates. What a business actually does to make its profit or how a life form actually behaves to preserve and propagate its genes represents the *proximate* means or mechanisms by which the ultimate goal is achieved.

Similarly, the ultimate function or goal of biological rhythms may well be to organize a complex system and align it with powerful geophysical cycles. This, too, is a critical and powerful insight. It allows us to anticipate the rhythmic nature of biological systems and expect to observe such rhythms at multiple levels of organization. However, it does not explain why gentamicin renal toxicity peaks during the inactive phase of the daily cycles in humans and other mammals. For that level of understanding, we need a different kind of explanation, a more proximate why. In fact, what we really need is a "how"—a mechanism. Because what we are really asking with the question "why does gentamicin toxicity peak in the rest phase?" is "How do daily rhythms in mammals result in a peak of renal toxicity in the rest phase?" Once we understand that mechanism, there is the possibility of taking advantage of it—in effect, getting the weeds out of the garden.

Figure 2.10 illustrates one such attempt to uncover the proximate mechanisms underlying a rhythmic phenomenon, in this case, the cycle of renal toxicity of gentamicin. In addition to a daily variation, it was noticed that toxicity was also related to food intake, with greater toxicity associated with reduced feeding. Beauchamp and Lebrecque[2] considered the following evidence:

1. There is a daily variation in renal toxicity with the highest toxicity associated with inactive phases in both laboratory animals and humans;

2. Renal toxicity also varies with food intake with reduction in feeding associated with increased toxicity;

3. As polyaminated molecules with a pK_a of between 6 and 9.5, aminoglycosides interactions with renal tubule cells will vary with urinary pH.

Putting this all together, they suggested the model I have illustrated in Figure 2.10 to explain the daily rhythm of gentamicin toxicity. In this model, the daily activity rhythms help generate feeding cycles which, in turn, drive rhythms in urinary pH which generate daily variation in renal toxicity. The advantage of having this model, if correct, is that practical applications can be created. For example, there are clearly better times of day at which to administer gentamicin to reduce side effects. However, even if that is not possible, the use of food or other means to increase urinary pH should reduce toxic effects on the kidney[2].

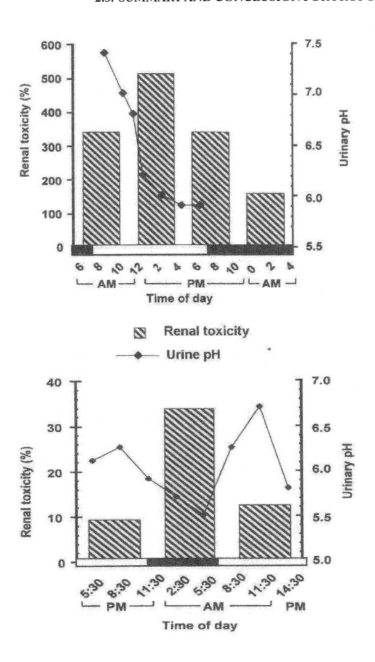

Figure 2.9: Diurnal variation in renal toxicity and urinary pH in laboratory animals (top graph) and human subjects (bottom graph). (From Beauchamp and Labrecque, 2007 with permission[2].)

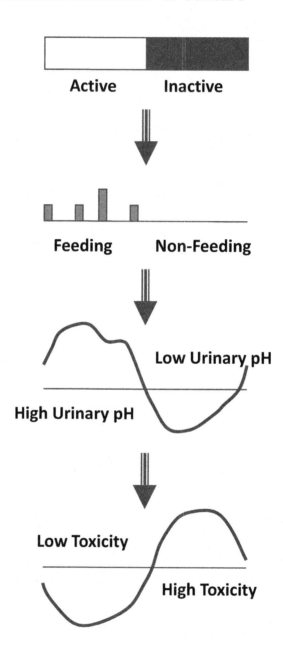

Figure 2.10: A possible model of the proximate mechanism underlying rhythmic variations in gentamicin toxicity.

Other proximate causal mechanisms can be suggested for other observed physiological and pharmacological cycles. For example, Figure 2.11 displays a rhythm in hepatic blood flow indicating a significant daily variation in liver perfusion[22]. This might well generate significant rhythms in the detoxification and/or bioactivation of blood borne chemicals (hormones, pharmaceutical agents, environmental toxins, etc.) regardless of whether or not the liver enzyme systems were also rhythmic simply by altering the availability of those chemicals to the liver cells. In Figure 2.12, significant daily cycles are displayed in two major classes of immune cells[32]. If the cells of the immune system vary by time of day, then the ability of the system to react—and extent of that reaction—can be expected to vary as well. Is it really any surprise, then, that disease symptoms display such prominent rhythms?

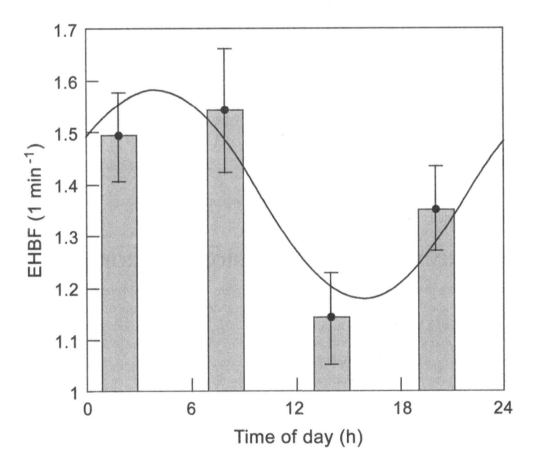

Figure 2.11: Daily variation in hepatic blood flow recorded in 10 healthy supine individuals using indocyanine green clearance. (From Lemmer, 1996 with permission[22].)

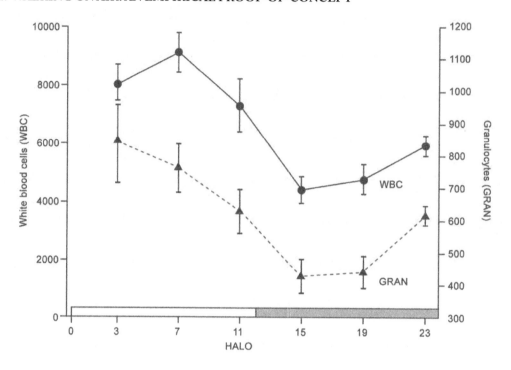

Figure 2.12:   Mean ± standard error for leukocyte and neutrophil cell counts in mice. (From Porsin, et al., 2003 with permission[32].)

2.4   SUMMARY AND CONCLUSION FOR PROPOSITION 2.2

The evidence that has been presented thus far confirms the viability of Proposition 2.2. The impact of environmental factors on biological systems does indeed vary with the time of exposure. We have not proven that all—or even a majority—of environmental factors depend on the time of exposure but rather that at least some do vary significantly in their impact across 24 hours. This is all that was necessary to establish a proof of concept. The extent of this variation, and the number of factors displaying such temporal dependencies, we leave until later.

Proposition 2.3   *Disruption of a biological rhythm will generate deleterious effects on the biological system.*

Before discussing this last and final proposition, it is necessary to carefully examine three important questions:

1. What is meant by the term "deleterious"?

2. What is the nature of the evidence that can be provided?

3. What is the underlying nature of daily rhythms?

2.4.1   QUESTION 1—WHAT IS MEANT BY THE TERM "DELETERIOUS"?

To determine if a change in rhythm status generates a deleterious impact, it is necessary to carefully define what is meant by a deleterious effect on a biological system. Recalling our discussion of ultimate and proximate function, a possible definition of deleterious could be anything that interferes with the biological system's ability to achieve its ultimate goal. The ultimate goal of all living things is the preservation and propagation of genetic material. Anything which interferes with that goal can thus be considered as deleterious. Since proximate mechanisms evolved to serve an organism's ultimate goal of gene propagation, and insofar as organisms themselves can be considered as proximate mechanisms, most deleterious effects can be assumed to be malfunctions harmful to the organism. In other words, if something is deleterious, it is probably damaging to the organism itself as well as to the gene's ultimate goal of propagation. However, this is not necessarily true. One biological rhythm humans often disrupt is that of the reproductive cycle in females. Disruption of this cycle is indeed deleterious to the evolved function of the system—disruption of the rhythm makes it far more difficult to get pregnant—but individual women disrupt it deliberately to achieve a positive benefit. Here the goal of the individual and the ultimate function of the genes come into conflict, and disruption of the system may be seen as a positive outcome from the individual's point of view. This personally relevant outcome should not blind us to the fact that—from the view of the ultimate function of the reproductive system—disruption of the biological rhythm is indeed deleterious to the purpose of the system which is to create another generation.

We already know of two biological rhythms where disturbances of normal cycling generates deleterious effects—the cardiac cycle and the menstrual cycle. In the former case, disruption is harmful on both a proximate (organismal) and ultimate (gene) level, whereas in the latter case, such disruption could be considered beneficial at the individual level while remaining harmful at the gene level. However, what can be said about disturbances in the daily rhythms we have been considering in this chapter? What are the effects of disrupting daily rhythms?

2.4.2   QUESTION 2—WHAT IS THE NATURE OF THE EVIDENCE?

If we wanted to use the most powerful scientific approach to the question of rhythm disorder, we would have to know how to experimentally disrupt daily rhythms. Otherwise, we are limited to weaker correlational studies where pre-existing disrupted rhythms are shown to correlate with some sort of physiological and/or behavioral abnormality. While such studies are useful, they are often limited in the applications that can be generated from them. In the case of smoking and lung disease discussed previously, correlational studies repeatedly associating smoking and lung damage were able to generate a fairly specific recommendation—do not smoke. If rhythms disruption were to show similar correlations between rhythm abnormalities and physiological and/or behavioral disorders, a similar prohibition might be suggested—do not disrupt your rhythms. However, unless we have some idea of how rhythms *can* be disrupted, this suggestion lacks much practical value. There is also the ever-present problem associated with correlational studies that lack experimental validation— the conclusion that rhythms play a causal role may simply be incorrect. If activity A leads to effect B and rhythm disruption simultaneously, then any physiological disorder associated with A could easily be the result of effect B and not altered daily cycles. Altering activity A to correct the rhythms without changing effect B will have little or no effect on the resulting disorder.

There is another reason for knowing something about how rhythms can be disrupted, even if correlational data were the sole evidence available. Knowing what factors create abnormalities in daily rhythms allows you to seek out those situations most likely to lead to such abnormalities and then ask what, if any, pathological changes are associated with those situations. If you had no idea what causes daily rhythms to deteriorate, then you would have no specific place to start looking. In effect, you would have to sample as many individuals with some kind of disorder as possible and look for rhythm abnormalities. Not only would this be fairly tedious, but one could never be sure whether rhythm changes caused the disorder or if the disorder led to rhythm changes.

2.4.3   QUESTION 3—WHAT IS THE UNDERLYING NATURE OF DAILY RHYTHM?

In order to understand what factors could disrupt daily rhythms we need to know something about how these rhythms are generated. There are two fundamentally different ways in which daily rhythms could be generated, both of which are compatible with the models and simulations discussed in Chapter 1. The simplest method would be to have one or more daily environmental cycles act as drivers, forcing a rhythm with a 24-hour frequency on the biological system(s). A second, more complex method would be to have the biological organisms evolve some sort of internal timer which would then have to be synchronized to the appropriate daily environmental cycle. Techniques to disrupt rhythms generated by these two methods would differ considerably depending on which— environmental driver or internal timer—the organism was using.

Up to this point, I have deliberately avoided using the term often supplied indiscriminately to all daily rhythms, *circadian*. There is a good reason for this. I believe that the study of biological rhythms would benefit from having precise terminology, and daily and circadian are not equivalent terms. A daily rhythm is any rhythm that shows a period of about 24 hours. Circadian rhythms are a subset of daily rhythms which are generated using an internal timer. It is fairly easy to tell the difference between examples of the two terms by placing the organism in question in *temporal isolation*. Temporal isolation is a condition in which there are no known environmental cycles. Non-circadian or environmentally driven daily rhythms will only persist for a few cycles before damping out of existence whereas true circadian rhythms will persist indefinitely.

Hopefully, the reason for my caution in using the term "daily" rather than "circadian" is clear to you. In order for a rhythm to be proved to be circadian, organisms displaying the rhythm in question will have to be placed in temporal isolation for many cycles. Indeed, Enright (1981) argued that proof that a 24-hour rhythm was truly circadian (according to the definition above) requires that a *phase reference point* on the biological rhythm scan the entire 24 hours of clock time. A phase reference point is any point to which the biological rhythm reliably returns. Examples include the maximum or peak in blood cortisol, the trough in body temperature, or activity onset in many animals. As an example, a daily rhythm with a period of 25 hours in temporal isolation will take 24 days to complete the scan and establish itself as a circadian rhythm. This is diagrammed in Figure 2.13 using an animal's rest/activity cycle as an example. Note that the phase reference point (activity onset) moves by 1 hour of clock time each cycle, reflecting the rhythm's period of 25 hours. Given the nature of the experimental conditions needed to confirm a daily rhythm as circadian, not all daily rhythms reported in the scientific literature will have undergone this rigorous evaluation and thus cannot be legitimately termed as circadian.

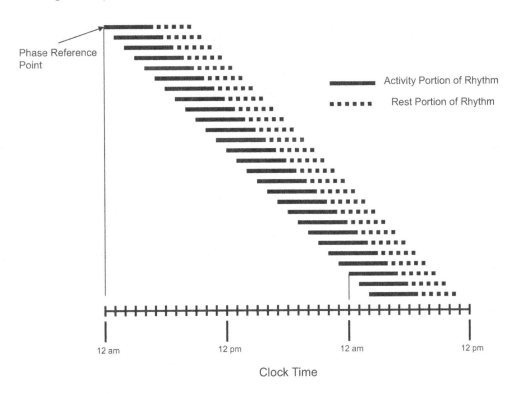

Figure 2.13: Representation of a circadian rest activity/cycle operating in temporal isolation.

How does the difference between environmentally driven daily rhythms and true circadian cycles influence the conditions needed to create rhythm disruptions? Rather profoundly in point of fact. An environmentally driven daily rhythm is, by definition, dependent upon an environmental cycle. Remove the cycle and within a few days, the rhythm will damp out and cease. Thus, it is fairly easy to create conditions that cause such rhythms to deteriorate by simply placing the organism under constant environmental conditions. On the other hand, if the rhythm is circadian, constant conditions may not cause any significant damage to a biological system. The observed cycle will not be exactly 24 hours long, but there would be no reason *a priori* to assume that such a slight change in frequency will cause any significant alteration in the organism's overall temporal coherence. So, how does one disrupt circadian rhythms? To answer this question, we need some sort of model or mechanism by which circadian rhythms can be generated.

Figure 2.14 provides the simplest possible model for a circadian system. There are five components in this model: 1. an environmental cycle or Zeitgeber (Ger. "time-giver") to which the circadian oscillator or pacemaker can synchronize; 2. a coupling mechanism (C_{zp}) which translates changes in the Zeitgeber into signals recognized by the pacemaker; 3. the circadian oscillator or pacemaker itself; 4. a coupling mechanism which translates changes in pacemaker activity into signals effecting physiological and behavioral outputs (C_{pr}); and 5. the output rhythms themselves.

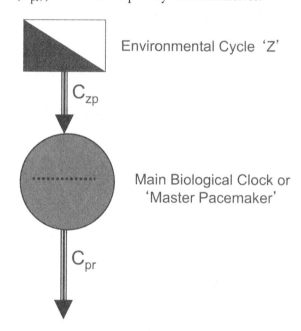

Figure 2.14: A diagram of the simplest possible circadian organization.

From the model, one can easily see why temporal isolation has a minimal effect on the system. Such isolation—also called constant environmental conditions—simply removes the incoming Zeitgeber signal. The rest of the system remains intact and as long as the other components operate within required parameters, the temporal coherence of the system should also remain intact.

Of course, this is the simplest model that could be imagined for a circadian system, and far more complex patterns could be envisioned. One possible variation would be the existence of secondary or "slave" oscillators[31]. Other variations include multiple pacemakers and coupling systems for a variety of different Zeitgebers.

At the moment, however, these circadian-based models are all hypothetical. I have provided neither argument nor data to determine which approach—environmentally driven daily cycles or circadian rhythms—is the normal means by which 24-hour biological cycles are generated in Earth's organisms. We have established the ubiquity of biological rhythmicity, but we have not yet determined the proximate mechanism(s) used to generate that rhythmicity. As a result, we are unable to determine how to predictably disrupt biological rhythms and are therefore not yet able to test the validity of Proposition 2.3. To truly examine Proposition 2.3, we will have to determine the mechanism(s) by which biological rhythms are generated in living species. That task begins in Chapter 3.

2.5   CHAPTER REVIEW

This chapter began with three propositions derived from the discussion of biological rhythms begun in Chapter 1. The original premises from Chapter 1 were:

1. Biological systems have a tendency to be rhythmic;

2. Biological systems will evolve to take advantage of this rhythmicity by organizing biochemical and physiological processes into temporal cycles;

3. Biological systems will evolve further to use at least some of these temporal cycles to adapt to stable geophysical cycles.

And these premises implied the following additional concepts:

4. Biological systems will then evolve timing mechanisms to serve as organizing interfaces between internal cycles and geophysical rhythms;

5. As a result of these evolutionary pressures, the biochemistry, physiology, and behavior of organisms will be organized into predictable temporal patterns.

   The propositions derived from these five ideas were:

a) Living systems will display significant rhythmicity;

b) The effect of environmental factors on living systems will vary based upon the time of exposure;

c) Disruption of biological rhythms will have a negative impact on the normal functioning of living systems.

   As a proof-of-concept in support of these propositions, evidence was collected and analyzed concerning one form of biological rhythmicity, daily rhythms linked to the 24-hour geophysical cycle of light and dark. Considerable evidence was provided in support of propositions (a) and (b) which, in turn, provides corroboration for the main premises from which those propositions were derived. In order to examine the third proposition, it will be necessary to determine the exact nature of daily biological rhythmicity. This journey begins in the next chapter, where we turn our attention to a more detailed analysis of the nature of biological rhythms.

REFERENCES

[1] Barton, N.H., Briggs, D.E.G., Eisen, J.A., Goldstein, D.B. and Patel, N.H. (2007). *Evolution*. Cold Spring Harbor Laboratory Press: Cold Spring Harbor, New York, pp. 109–136. 37

[2] Beauchamp. D. and Labrecque, G. (2007). Chronobiology and chrontoxicology of antibiotics and aminoglycosides. *Advanced Drug Delivery Reviews* 59(9–10): 896–903. DOI: 10.1016/j.addr.2006.07.028 45, 48, 49

[3] Brady, J. (1981). Behavioral rhythms in invertebrates. In Aschoff, J. (ed.) *Handbook of Behavioral Neurobiology 4: Biological Rhythms*. Plenum Press: New York, pp. 125–144. 35

[4] Constance, C.M., Fan, J.-Y., Preuss, F., Green, C.B., and Price, J.L. (2005). The circadian clock-containing photoreceptor cells in *Xenopus laevis* express several isoforms of casien kinase I. *Molecular Brain Research* 136: 199–211. DOI: 10.1016/j.addr.2006.07.028 38

[5] Christina, A.J.M., Merlin N.J., Vijaya, C., Jayaprakash, S. and Murugesh, N. (2004). Daily rhythm in nociception in rats. *Journal of Circadian Rhythms* 2. http://www.jcircadianrhythms.com/content/2/1/2 DOI: 10.1186/1740-3391-2-2

[6] Cutolo, M. and Straub, R.H. (2008). Circadian rhythms in arthritis: Hormonal effects on the immune/inflammation reaction. *Autoimmunity Reviews* 7(3): 223–228. DOI: 10.1016/j.autrev.2007.11.019 39

[7] Enright, J.T. (1981). Methodology. In Aschoff, J. (ed.) *Handbook of Behavioral Neurobiology 4: Biological Rhythms*. Plenum Press: New York, pp. 11–19.

[8] Eriksson, M.E. and Millar, A.J. (2003). The circadian clock: A plant's best friend in a spinning world. *Plant Physiology* 132: 732–738. DOI: 10.1104/pp.103.022343

[9] Fan, J.-Y., Muskus, M.J., Preuss, F. and Price, J.L. (2008). Evolutionarily conserved features of vertebrate chi delta and *Drosophia* Dbt in the circadian mechanism. Presented at the Society for Research on Biological Rhythms 20th Anniversary Meeting, Destin, FL May 17–21. Abstract 37, Pg 80. 38

[10] Freedman, D., Purves, R. and Pisani, R. (1998). *Statistics*. W.W. Norton & Company: New York. 32

[11] Golden, S.S., Ishiura, M., Johnson, C.H. and Konda, T. (1997). Cyanobacteria circadian rhythms. *Annual Review of Plant Physiology and Plant Molecular Biology* 48: 327–354. DOI: 10.1146/annurev.arplant.48.1.327 35

[12] Halberg, F., Johnson, E.A., Brown, B.W. and Bittner, J.J. (1960). Susceptibility rhythm to E. coli endotoxin and bioassay. *Proceedings of the Society for Experimental Biology* (N.Y.) 103: 142–144. DOI: 10.3181/00379727-103-25439 44, 142

[13] Harrison, H. (1960). *The Deathworld Trilogy*. Nelson Doubleday: Garden City, New York. 29

[14] Haus, E., Halberg, F., Kuhl, J.F.W. and Lakatua, D.J. (1974). Chronopharmacology in animals. In Aschoff, J., Ceresa, F. and Halberg, F. (eds). *Chronobiological Aspects of Endocrinology* (Symposia Medica Hoechst 9). F.K. Schattauer Verlag: Stuttgart-New York, pp. 269–304. 44, 45, 46

[15] Hecht, S.S. (2005). Carcinogenicity studies of inhaled cigarette smoke in laboratory animals: old and new. *Carcinogenesis* 26: 1488–1492. DOI: 10.1093/carcin/bgi148 34

[16] Hempel, C.G. (1965). *Aspects of Scientific Explanation*. The Free Press: New York. 32

[17] Horton, T.H. (2001). Conceptual issues in the ecology and evolution of circadian rhythms. In Takahashi, J.S., Turek, F.W. and Moore, R.Y. (eds.) *Handbook of Behavioral Neurobiology 12: Circadian Clocks*. Kluwer Academic/Plenum Publishers: New York, pp. 45–57. 38

[18] Hutt, J.A., Vuillemenot, B.R., Barr, E.B., Grimes, M.J., Hahn, F.F., Hobbs, C.H., March, T.H., Gigliotti, A.P., Seilkop, S.K., Finch, G.L., Mauderly, J.L. and Belinsky, S.A. (2005). Life-span inhalation exposure to mainstream cigarette smoke induces lung cancer in B6C3F1 mice through genetic and epigenetic pathways. *Carcinogenesis* 26: 1999–2009. DOI: 10.1093/carcin/bgi150 34

[19] Johnson, C.H. (2001). Endogenous timekeeping in photosynthetic organisms. *Annual Review of Physiology* 63: 695–728. DOI: 10.1146/annurev.physiol.63.1.695

[20] Johnson, C.H. and Golden, S.S. (1999). Circadian programs in cyanobacteria: Adaptiveness and mechanism. *Annual Review of Microbiology* 53: 389–409. DOI: 10.1146/annurev.micro.53.1.389 35

[21] Lakin-Thomas, P.L. and Brody, S. (2004). Circadian rhythms in microorganisms: New complexities. *Annual Review of Microbiology* 58: 489–519. DOI: 10.1146/annurev.micro.58.030603.123744 35, 38

[22] Lemmer, B. (1996). The clinical relevance of chronopharmacology in therapeutics. *Pharmacological Research* 33(2): 107–15. DOI: 10.1006/phrs.1996.0016 51

[23] Lyttleton, R.A. (1982). The nature of knowledge. In Duncan, R. and Weston-Smith, M. *The Encyclopedia of Ignorance*. Pergammon Press: New York, pp. 9–17. 32

[24] McClung, C.R. (2001). Circadian rhythms in plants. *Annual Review of Plant Physiology and Plant Molecular Biology* 52: 139–162. DOI: 10.1146/annurev.arplant.52.1.139 35

[25] Medawar, P. (1984). *The Limits of Science*. Oxford University Press: Oxford, Eng. 32

[26] Millar, A.J. (2004). Input signals to the plan circadian clock. *Journal of Experimental Botany* 55(395): 277–283. DOI: 10.1093/jxb/erh034 35

[27] Moore-Ede, M.C., Sulzman, F.M. and Fuller, C.A. (1982). *The Clocks that Time Us*. Harvard University Press: Cambridge, MA.

[28] Murtas, G. and Millar, A.J. (2000). How plants tell time. *Current Opinions in Plant Biology* 3(1): 43–46. DOI: 10.1016/S1369-5266(99)00034-5

[29] Parascandola, M. (2004). Skepticism, statistical methods, and the cigarette: A historical analysis of a methodological debate. *Perspectives in Biology and Medicine* 47(2): 244–261. DOI: 10.1353/pbm.2004.0032 34

[30] Pariat, C., Courtois, P., Cambar, A., Piriou, A. and Bouquet, S. (1988). Circadian variations in the renal toxicity of gentamicin. *Toxicology Letters* 40(2): 175–182. DOI: 10.1016/0378-4274(88)90159-2 44, 47

[31] Pittendrigh, C.S. (1981). Circadian rhythms: General perspective. In Aschoff, J. (ed.) *Handbook of Behavioral Neurobiology 4: Biological Rhythms*. Plenum Press: New York, pp. 57–80. 57

[32] Porsin, B., Formenta, J.L., Filipski, E., Etienne, M.C., Francoual, F., Renee, N., Magne, N., Levi, F. and Milano, G. (2003). Dihydropyrimidine dehydrogenase circadian rhythm in mouse liver: Comparison between enzyme activity and gene expression. *European Journal of Cancer* 39(6): 822–828. DOI: 10.1016/S0959-8049(02)00598-1 51, 52

[33] Root-Bernstein, R. and McEachron, D.L. (1982). Teaching theories: The evolution-creation controversy. *American Biology Teacher* 44: 413–420, 1982. Article reprinted in: *Anthropology 83/84* (Elvio Angeloni, ed.) The Duskin Publishing Group Inc., Guilford, CT, pp.162–168. 32

[34] Rosbash, M. (1995). Molecular control of circadian rhythms. *Current Opinion in Genetics and Development* 5: 662–668. DOI: 10.1016/0959-437X(95)80037-9

[35] Rusak, B. (1981). Vertebrate behavioral rhythms. In Aschoff, J. (ed.) *Handbook of Behavioral Neurobiology 4: Biological Rhythms*. Plenum Press: New York, pp. 183–213. 35

[36] Smolensky, M., Reinberg, A. and Labrecque, G. (1995). Twenty-four hour patterns in symptom intensity of viral and allergic rhinitis: Treatment implications. *Journal of Allergy and Clinical Immunology* 95(5): 1084–1096. DOI: 10.1016/S0091-6749(95)70212-1 41, 42, 43

[37] Sweeney, B.M. (1963). Biological clocks in plants. *Annual Review of Plant Physiology* 14: 411–440. DOI: 10.1146/annurev.pp.14.060163.002211 35

[38] Takahashi, J.S., Turek, F.W. and Moore, R.Y. (2001). *Circadian Clocks*. Kluwer/Plenum: New York. 35

[39] Vlahos, R., Bozinovski, S., Jones, J.E., Powell, J., Gras, J., Lilja, A., Hansen, M.J., Gualano, R.C., Irving, L. and Anderson, G.P. (2006). Differential protease, innate immunity, and NF-kB induction profiles during lung inflammation induced by subchronic cigarette smoke exposure in mice. *American Journal of Physiology—Lung Cellular and Molecular Physiology* 290: L931–L945. DOI: 10.1152/ajplung.00201.2005 34

[40] Wager-Smith, K. and Kay, S.A. (2000). Circadian rhythm genetics: from flies to mice to humans. *Nature Genetics* 26: 23–27. DOI: 10.1038/79134 38

CHAPTER 3

Clock Tech, Part 1

Overview

In this chapter, we begin our search for the mechanisms underlying biological rhythms. In Chapter 1, we determined that there were at least two fundamental adaptive reasons for the evolution of biological rhythms. The first reason was that all complex goal-driven devices must time their activities in order to maximize efficiency. The second reason lay in the rhythmic nature of the Earth's environment and the need for organisms to adapt to geophysical cycles. These twin requirements—*internal* and *external temporal order*—must both be satisfied by whatever mechanism has evolved to *control timing* in living systems. In Chapter 2, we were able to establish that such mechanisms have evolved, at least for daily rhythms. We now investigate whether or not *environmentally driven rhythms* are sufficient to satisfy these two requirements. First, we will examine the situation if each internal rhythm is driven by its own unique environmental cycle. The result of our analysis will indicate that such a system is unlikely to maintain internal temporal order. Next, we will examine the possibility that using a *single environmental cycle*—the day/night cycle—to drive all internal rhythms will be able to satisfy the requirement for internal temporal order. In the process, we will expand our analysis to a consideration of *seasonal cycles* and the need for the mechanisms underlying biological rhythms to *adapt to multiple geophysical cycles*. We will also examine the nature of living systems and the relationships between organisms and their environment in light of biological evolution.

To be interested in the changing seasons is a happier state of mind than to be hopelessly in love with spring.

-George Santayana

3.1 EVOLUTION OF A MECHANISM

In this chapter, we begin discussing what mechanisms could have evolved to generate and regulate rhythms within a biological organism. The term "mechanism" is used in a general engineering sense, less in terms of the actual physical machinery and more like a design or blueprint. In other words, how might the biological machinery have evolved to cope with naturally occurring geophysical cycles while maintaining the necessary internal temporal order?

To approach this question properly, we should define the nature of the initiating stimulus, X, in the system. Referring to Figure 3.1, the exact nature of stimulus "X" is somewhat ambiguous.

For example, X could indicate cold and be considered an external, environmental factor or could represent a decrease in blood glucose levels and be considered an internal, physiological factor. Does this distinction have any real impact on our analysis? As we will find as we continue to explore biological rhythms, the answer is both *yes* and *no*. Certainly there is a real difference between an environmental factor and a physiological/biochemical one. Organisms, and for that matter cells and even subcellular organelles, can be modeled as thermodynamic systems with the rest of the universe acting as surroundings. In such models, an X factor as part of the surroundings is quite different from an X factor acting as part of the system itself. So, from a chemical and/or physical point of view, there is a real difference. The trick is to correctly identify the system in question and thus the proper boundaries between the system and its surroundings.

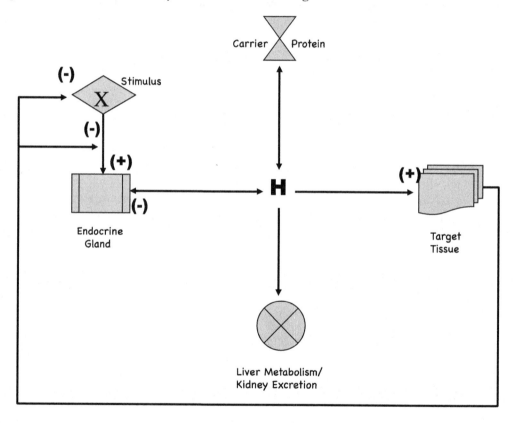

Figure 3.1: A simple endocrine model system.

From an evolutionary point of view, however, this dichotomy between external and internal factors appears less clear-cut. On the one hand, biological evolution never occurs in isolation, but rather within an actual environment with which individuals must cope in order to pass their genes successfully on to subsequent generations. From that perspective, the external environment could be considered part of the evolving systems. The X shown in Figure 3.1 would not then be an external factor. The so-called "external" environment is so intimately involved with the evolution of biological systems that it could be considered an element of those systems.

The term that is used when considering the inter-relationship of species is *co-evolution*[2]. A parasite has a certain infective mechanism to which the host develops a response. This leads to selection for a different infective mechanism which prompts a different response, etc. However, perhaps more dramatically, even the physical environment is altered by living systems that are, in turn, changed by the environment. Free atmospheric oxygen is actually a byproduct of photosynthesis and ozone a byproduct of free oxygen. Thus, the evolution of photosynthetic organisms led to increasing atmospheric oxygen, which in turn allowed for the colonization of the land, which then allowed for the evolution of new species of photosynthetic organisms[6]. The concept that the Earth is a single, complex evolving system thus has a certain appeal and even validity.

On the other hand, information about the external environment often must be changed into a form that is useful to the biological system or organism, much as a room's ambient temperature must be transduced into an electrical signal in a thermostat. In such cases, it is not the actual information from the external environment that influences the system but rather the transformed information. The act of taking environmental stimuli and transforming those stimuli into a form that can be utilized by a biological system is called *transduction* and is never perfect. Thus, what an organism perceives about its environment and the actual state of the environment are always slightly different. Evolutionary processes act to minimize this difference but it can never be entirely eliminated. When I was growing up, a popular, albeit illegal, recreational drug was LSD—lysergic acid diethlamide. It was known to alter the user's perceptions about the world. Of course, the world did not change but the user's perceptions often did. This is possible because the drug altered the manner in which information about the environment reached the processing centers of the brain. This demonstrates the importance of properly transducing and processing environmental information if that information is to be correctly utilized to promote an organism's evolutionary success. From this perspective, it would be best to think about X as an external factor since doing so highlights the critical importance of correct transduction and processing.

So which particular view of X is correct? In one sense, all are correct views. Which approach is most useful depends on the circumstances by which you define the system under study. In an ecosystem, many environmental factors external to the individual organism would be included as part of the overall system. If you are studying a cellular organelle, such as a mitochondrion, factors we would normally think of as internal to the individual, such as serum glucose level, would have to be viewed as external factors relative to the organelle. In short, there is not just one system or one level. Living things operate within a multitude of systems operating at many different levels at scales ranging from molecules within cells to the planet as a whole. The important thing when considering models, such as those we have been discussing so far, is to correctly identify the system and level at which that system is operating.

Why did I dwell on this admittedly somewhat abstract point to such an extent? Later on, we will find the manner in which a biological system is defined makes a significant practical difference in how that system is analyzed and modeled in terms of rhythms. For now, let's simply consider the implications of systems and evolutionary thinking on how physiology really works. Most likely, you think of yourself as a single individual and separate system, although probably not in those terms. The very word of "self" is a unitary concept reinforcing that perspective. However, this perspective is too limited for a full understanding of biological rhythms. Looking inward, this "self-system" outlook inhibits us from recognizing that humans are actually vast collection of interacting subsystems which are connected in various ways and therefore have the potential to become disconnected. This potential to disconnect various internal systems—and the need to prevent this from occurring—has important implications for the functioning of biological rhythms.

Looking outward from the individual, the separation of self from the external world leads humans to underestimate the biological impact of the external environment. When moving into new environments, different from those in which they evolved, humans use technology as a buffer against adverse conditions. We are so successful at this that we have begun to consider ourselves as practically infinitely adaptable. What other species can travel to the moon and back and survive? However, the fact is that we do not adapt to new environments as much we use technological means to package up the necessary components of the old ones and take them with us. We do not, after all, go into space naked—we take the necessary heat and air with us. In fact, whenever we do not take those elements to which we have evolved along, there tend to be negative consequences. One of the results of long-term space travel is muscle atrophy, bone loss, and other physiological deficits, because we have yet to develop a practical method of taking Earth's gravity with us[1,4,5,7,8,9]. These negative consequences are the result of losing a vital component of living systems—an X factor (in this case, Earth's gravity)—and the human physiological system degrades as a result. If this is true of gravity, could it not be true also of geophysical cycles? What happens when time is the X factor? To answer that last question, we will explore how biological systems evolved to incorporate the X of geophysical cycles.

3.1.1   ENVIRONMENTALLY DRIVEN RHYTHMS

The first potential mechanism capable of generating and regulating biological rhythms is also the most simple, basic, and easy to understand. If X is a geophysical cycle and is thus rhythmic, and X is a component of the system, then the entire system—including its biological elements—will become rhythmic. No specific changes are required to the model provided in Figure 3.1. In fact, this was already established by Mr. Freedman's mathematical simulation, the results of which are reproduced again in Figure 3.2. Clearly, if the input stimulus to the physiological system oscillates, the entire system oscillates in response. No further physiological adjustments or mechanisms are needed. Even a single stimulus generates a temporary oscillation, as seen again in Figure 3.3. So, as long as X cycles, the entire system will cycle with the same frequency. Problem solved—or is it? What would occur if X were to be changed so as to be a steady, rather than oscillating, input?

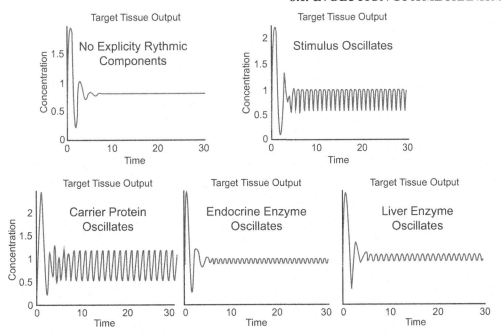

Figure 3.2: Target tissue response plots showing the effects of when a single component of system becomes rhythmic.

At first, that seems like a rather dumb question. How could geophysical cycles stop cycling? Wasn't the whole point of adapting to those environmental rhythms based on the rhythms' extreme stability? The answer lies in the nature of the linkage or coupling between factor X and the physiological response to X. In this case, we will consider X as external to the biological system. Outside of certain biblical tales (see Joshua 10:12–14), geophysical cycles are not going to disappear. However, it is possible an organism's ability to sense those cycles might temporarily vanish. For example, if an animal sleeps in a cave or burrow away from sunlight, it is a valid question to ask how that organism could sense the environmental rhythm of light and dark. If biological rhythms are being directly generated through environmental cycles stimulating internal rhythmicity, it does not matter if those cycles exist if the organism has no method of *sensing* them. Remember, if X cannot be perceived, it is same as if X did not exist at all. So it really is a valid question—what happens to rhythms in organisms when those organisms become isolated from the environmental cycle driving those rhythms, even for a short period of time?

From our model, it appears that all is not lost if the isolation from environmental cycles does not last for too long a period of time. Focusing solely on the temporal aspects of the system, rhythms may persist or continue for a while and then damp out and become stable at a single operational level within the ability of the system to maintain that stability. Why would rhythms persist at all without

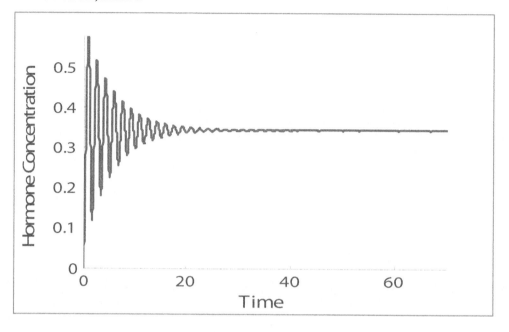

Figure 3.3: Circulating hormone concentration in the bloodstream displays a damped oscillation in response to single input stimulus.

X cycling? Consider that the system beyond X relies on numerous feedback loops (Figure 3.1). Once started, these loops tend to be oscillatory. So, when X is rhythmic, H becomes rhythmic, and its effects on its own secretion and on the target are rhythmic and so on. Assuming an initial cycling of X to start the system, when X eventually becomes constant, these oscillatory loops are already in operation. Since H is rhythmic, it forces a rhythm on other components of the system which, in turn, keep H rhythmic and so on. Whether the rhythm would damp out or persist even in the face of a constant level of X depends greatly on how X affects the system. In Figure 3.1, X stimulates the secretion of H and thus competes with the negative feedback of H itself and the output of the target tissue. If X has a greater effect than H and the target tissue output, this stimulation would overcome the negative effects of H and target tissue, thus forcing high levels of H at all time points. This would then abolish the rhythm throughout the system. If the effect of X is less than the effects of H and the target tissue, the amplitude of the rhythm would decrease but rhythm itself might persist for many cycles. Thus, how long rhythms would persist in the system stimulated by constant levels of X would depend on the positive effects of X on H secretion relative to the negative effects from H and the target tissue.

Before we get too caught up in specific biological details, note that while I continue to use the endocrine-based model shown in Figure 3.1 to describe how various timing mechanisms could evolve, nothing about these evolutionary approaches is limited to endocrine systems. A similar kind of analysis could be applied to any biochemical and/or physiology systems with negative feedback as a controlling element.

In this mechanism, then, X is the driver for the system. It is an oscillating system only because X is cycling. If X reflects or is linked to a geophysical cycle, then the frequency of the biological rhythm will match that of the geophysical oscillation. This category of biological rhythm is called *exogenous* or *exogenously generated,* reflecting the dependence of the rhythm on a cycle external to the biological components. Perhaps the reader can now see the reasoning behind discussing the role of X as a component of the system. This is a rhythmic system if and only if X is rhythmic and part of the overall system. Remove the oscillating X component, either by making X non-rhythmic or by eliminating the connection between X and the other components—and the system no longer cycles.

3.1.2   ARE ENVIRONMENTALLY DRIVEN RHYTHMS SUFFICIENT?

Clearly, then, one way to generate and regulate biological rhythms is to directly link the internal elements with the external cycles. In effect, we are expanding our definition of the system to include environmental components. While relatively straight forward, this kind of approach is highly dependent on those external cues to coordinate rhythmic organization. In addition, the way we set up this model has each biological rhythm dependent on its own specific environmental cues without any type of centralized organization. For example, a thyroid hormone might respond to environmental temperature cycles, behavior (recalling our on-time bird) could depend on light/dark cycles and other physiological rhythms on yet other environmental rhythms. As we will discuss later on, this could lead to considerable disorganization, since some environmental cycles—such as temperature—may be more variable than others, such as day/night. Since internal temporal order is a necessary adaptation for complex, goal-driven systems such as life forms, this approach seems a bit too risky to be the sole means of regulating biological rhythms. Some sort of limiting, central control mechanism would seem required to minimize the danger of internal temporal disorder.

This approach of multiple environmental cycles driving multiple internal rhythms can create problems in external, as well as internal, temporal order. Recalling the on-time bird from Chapter 1, suppose our feathered friend were to use a cycle of cold and hot instead of light and dark to determine the right time to appear on the feeding grounds. For example, the bird could use a three-hour delay from a certain level of cold to know when to seek out breakfast. But what if there is a warm snap one particular morning and it does not reach the required level of cold until an hour after normal? This will result in a change in the phase of the X oscillation which might, in turn, change the phase of the behavioral cycle. But, while such a change in air temperature would alter our bird's temporal system, it may not be enough to alter the behavior of the worms. As such, a brief change in air temperature could then result in one very hungry bird—and several happy worms, of course.

Let's consider another example, one that does not rely on a hungry bird or even a daily cycle. Certain types of mammals living in the northern reaches of the United States and Canada undergo seasonal changes in fur color, brown or grey during the summer and white in the winter[3]. How should the strategy underlying this change have evolved? The use of environmental cycles to drive biological rhythms is basically a reactive mechanism. If we consider a hare, a reactive change might mean that the physiology controlling fur color would begin to alter the color of the fur only after snow begins to appear in the animal's environment. As a result, the animal would remain brown or grey in a predominantly white environment for as long it took to alter the fur color. This would be the equivalent of painting a bull's-eye on the poor creature for the local predators.

On the other hand, suppose the hare evolved a different approach which allowed the animal to anticipate or predict the new winter conditions using the geophysical cycle, such as photoperiod change, that indicated time of year. This would avoid the bulls-eye issue by changing fur color ahead of time if and only if it snows on schedule. Otherwise, the poor hare will be white in a predominantly brown environment and no better off than before.

So which is best? As is often the case with evolutionary questions, the answer is: it depends. How predictable is the appearance of snow? How long does it take for the average hare to change fur color from brown/grey to white? What is the predation risk by being colored incorrectly for the environment? All these factors, and many more, form part of an evolutionary cost/benefit analysis. This complex ratio generates the selection pressure to evolve a mechanism that provides the best possible solution balancing prediction and reaction.

Note that the issue is not about whether or not the hare in question will display a rhythm—annual, in this case—in fur color. That will occur regardless of whether the animal uses a reactive or predictive mechanism. The real question is whether the animal passively waits for a direct stimulus—the appearance of snow—to signal the time to change fur color (reactive) or actively anticipates the likely future appearance of snow using some alternative timing system (predictive).

Ultimately, the best strategy will depend on the stability of future events. This might seem a bit counterintuitive at first: future events are by their very nature unreliable. However, it is a reasonable bet that the Earth's rotation will generate a sunrise tomorrow, and so planning on being awake in the morning to take advantage of daylight is an activity with a high probability of a positive payoff. On the other hand, the probability of no rain tomorrow is quite a bit less than that of sunrise—at least in Philadelphia—so I may regret not taking an umbrella to work. So, the problem being addressed through evolutionary processes is how best to deal with the situation of temporal events—such as worm availability or the appearance of snow—linked to, but not as stable or predictable as, major geophysical cycles.

It appears that we have two related issues complicating the simple model of environmental cycles driving biological rhythms. The first issue involves the internal and external coordination of rhythms. If each biological rhythm is stimulated by a specific environmental cycle related to the functioning of the particular rhythm, the result will be a variety of environmental cycles driving a similar variety of biological rhythms. This creates a situation where there is considerable risk of internal and external temporal disorder.

The second issue involves the adaptive function of rhythms in preparing organisms for predictable environmental changes, such as sunrise or seasonal snow cover. The mechanism of having environmental cycles driving biological rhythms is primarily reactive rather than predictive. A change in the environmental conditions produces a change in an organism's internal physiology and if the environmental conditions cycle, so will the physiological circumstances. This makes it difficult to anticipate and prepare for environmental changes, a key adaptive advantage as discussed with the on-time bird and colorful hare.

Thus, while providing some measure of temporal organization, a model based upon multiple environmental cycles driving multiple biological rhythms does not appear to pass evolutionary muster as a sufficient adaptation.

3.1.3   LIMITATIONS WITH ENVIRONMENTALLY DRIVEN RHYTHMS AND A NEW MODEL

One method for correcting the problems raised in the preceding section would be to limit the number of environmental stimuli driving the biological rhythms. Use a few very stable environmental cycles—such as the day/night cycle—to drive most, if not all, internal biological rhythms with similar ($\sim$ 24 hour) periods. Thus, any biological rhythm with a daily cycle would be driven by the environmental light/dark (LD) cycle and no other environmental rhythm. Once the overall daily biological rhythms are stable, a system could evolve to coordinate the activity of rhythms with different frequencies into an overall temporal organization. How might this evolve?

To keep it simple, let's limit the discussion to daily rhythms. One possibility would be to have each internal biological rhythm linked directly to the external LD cycle. This would require that each rhythm system evolve a specific LD sensor. This was not needed previously since each rhythm was responding to its own environmental stimulus. For example, if the stimulus were a decrease in body temperature, then the result was secretion of thyroid hormone; if the stimulus were a drop in serum glucose, the result might be secretion of cortisol and so on. Each system responded to its own stimulus which was directly related to the action of that physiological system—thyroid hormones increase body temperature; cortisol increases serum glucose; etc. Now, we have added a new wrinkle—these physiological systems should respond to light/dark cycles as well as to their own specific stimuli.

An alternative approach would be to evolve a centralized LD sensing structure and distribute the information to the various physiological systems that need to regulate daily rhythmicity. This would reduce considerably the number of sensors needed to link internal rhythms with external cycles. In this case, the physiological rhythms are indirectly linked to the external LD cycle via this central LD sensing structure.

So which is best and most likely to evolve? There are costs and benefits to both approaches. The individualized LD sensor approach has organizational redundancy so that if a single LD sensor were to fail, only the rhythm controlled by that specific sensor would be affected—the rest would continue on. On the other hand, multiple sensors means multiple sensitivities—different sensors will likely react slightly differently to the LD cycle. As a result, the timing information provided by the sensors could vary, resulting in slightly different timing among the various physiological rhythms. This is the very problem we are trying to avoid. In addition, there is the problem of sensor location—

light may not penetrate to the gland or organ responsible for the physiological system. This would require the evolution of multiple remote sensor systems, which is possible but seems unnecessarily complex.

On balance, it appears that a more centralized approach makes the best evolutionary sense. The fact is, however, that we cannot really establish this result based upon argument alone and need empirical evidence. For the time being, let's assume that a limited number of central LD receptors evolve and examine the implications of this solution to the problems associated with the multiple environmental drivers. If evidence from organisms does not support this approach, we can revise it later on.

So, how does this change the model we have been working with up to now? In Figure 3.4, I have added the significant new elements in red and green (tomorrow is Christmas). The new elements include a stable geophysical cycle—the LD cycle—a sensor for that cycle, and a method by which changes in sensor status can impact the physiological system. The original stimulus, X, has been subdivided into two components—a direct stimulus, X_{Dir}, and an integrated stimulus, X_{Int}. The X_{Dir} term is equivalent to the original stimulus, X, and represents any environmental change, internal or external, which stimulates the system to respond. The X_{Int} term is new and has been added to integrate inputs from both direct stimulation and the sensor linked to the geophysical cycle. Thus, the output, H, is now dependent on both direct stimulation and temporal information related to the light/dark cycle. The resulting rhythm is stabilized through its linkage to the very stable geophysical cycle of day and night.

Figure 3.4 does not differentiate between each rhythm having its own, individual LD sensor system and a more centralized approach with a limited number of LD sensors controlling multiple rhythms. The latter approach is diagrammed in Figure 3.5.

3.1.4 MULTIPLE RHYTHMS AND PHASE RELATIONSHIPS

Using both argument and models, we uncovered certain limitations with the simplest method of generating biological rhythms with the same frequency as important geophysical cycles. If each internal rhythm linked to its own specific environmental cycle, slight variations in the timing of different environmental stimuli could lead to both internal and external temporal disorder. Internal disorder is analogous to a car out of tune or a computer with a defective timing system—the car or computer runs inefficiently or breaks down completely under those circumstances. External disorder means that the organism is not fitting in to the external temporal environment—a brown hare running around in a white landscape or a bird arriving too early or too late on the feeding grounds. In a way, the ecosystem would be out-of-tune as a result. This external disorder could be magnified by another limitation of the multiple environmental drivers-multiple internal rhythms model. The model is primarily reactive rather than predictive. Internal rhythms are responding only to changes in stimuli directly related to their physiological functions—there is no anticipation of predictable future changes, only reaction to changes as they occur. In Chapter 1, it was argued that biological rhythms evolved in part as an adaptation to geophysical cycles in order to allow organisms to predict

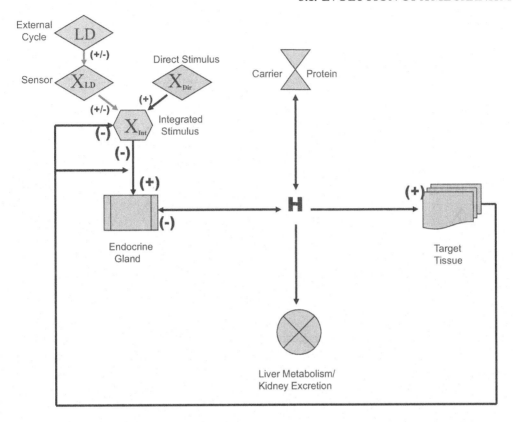

Figure 3.4: Simple endocrine model system revised through linkage with external light/dark cycle.

stable temporal changes in their environments. That was the whole point of discussing the on-time bird. Unfortunately, the simplest model as presented in Figure 3.1 does not accomplish this task very well. The model does show how inherently rhythmic physiological feedback systems can be rhythmic and how an environmental cycle might drive a corresponding biological rhythm. That said, the model does not work well for coordinating multiple internal rhythms or for generating predictions in a world dominated by stable geophysical cycles. Thus, it was necessary to modify the model with the new model(s) presented in Figures 3.4 and 3.5.

   In the new models, the number of environmental cycles capable of driving corresponding biological rhythms was reduced. In fact, it was reduced to just one—the day/night or light/dark (LD) cycle. This may turn out to be too extreme, but for the time being let us see how this modification affects the limitations of the original model. Again, referring to Figure 3.1, if X is a geophysical oscillation driving—or at least initializing—the system, the only result we have considered is that the biological components of the system will become rhythmic in response. It was assumed in

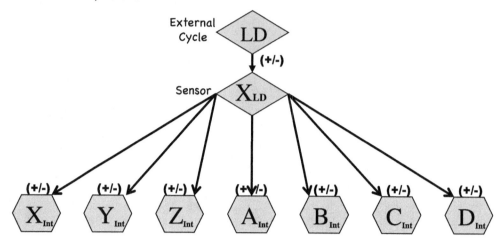

Figure 3.5: Rhythmic control using centralized sensor for light/dark (LD) cycles.

this analysis that the frequency of X would determine the frequency of the entire system. One improvement in the new model is that all biological rhythms are being driven by one environmental cycle, assuring that all internal rhythms have the same frequency. However, rhythms have other characteristics beyond that of frequency. Three other essential characteristics are the amplitude, the range, and the phase (see Figure 3.6). Of these three, it is the phase that we will consider now because it is phase that determines the predictability of a biological rhythm in anticipating, rather than reacting, to environmental changes.

The phase of the rhythm determines how the pattern of the oscillation fits into a specific frequency. For example, consider Figure 3.7. All three rhythms have exactly the same frequency—in fact, they are simply copies of a single oscillation! However, they all have different phases, relative both to each other as well as to clock time, which is represented by the line segment at the bottom of the figure. In this case, clock time represents the geophysical cycle of day and night, but it could refer to any environmental rhythm. Thus, numerous rhythms can exist with the same frequency but with different phases.

Returning to Figure 3.1, what is the situation in terms of the relative phases of the oscillations of X and H? The simple answer is: it depends. Depends on what? Upon the time it takes for changes in X to generate changes in the secretion of H. For example, if a change in the intensity or concentration of X instantly alters the secretion of H, then the oscillations in X and H will occur at the same time relative to each other (Figure 3.8a). It is said in this case that the two rhythms are *in phase* with each other. Alternatively, suppose that there was an hour delay between a change in X and a change in the secretion of H. This situation is shown in Figure 3.8b. In this latter example, it would be said that there is a *phase angle* of 1 hour between the oscillations of X and H. Thus, the

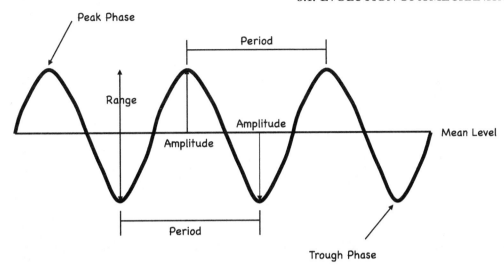

Figure 3.6: General characteristics of a rhythm.

particular temporal relationship between changes in X and changes in H determines the phase angle between them. This is sometimes referred to as the *coupling* between the X and H rhythms.

If we consider the situation within a biological organism, we begin to see the limitations imposed upon our thinking by the model provided in Figure 3.1. This model—no matter how useful—shows only one biological subsystem within an organism. Organisms can have dozens, hundreds, maybe even thousands of such systems. How likely is it that all such rhythms would have the exact same phase? In other words, would all rhythms within an organism have a relative phase angle of 0?

The answer would seem clear—not likely at all. Since different subsystems have different functions, some of which are opposed to each other, it seems impossible that all rhythms would have the same relative phase to each other and to environmental geophysical cycles. As an example, simply consider a single enzymatic system in the human liver, where hepatic phosphorylase breaks down glycogen into glucose and glycogen synthetase acts on glucose to synthesize glycogen. If both enzyme systems were to increase and decrease at the exact same times, the result would be complete stasis—no change in glucose levels would ever happen (Figure 3.9). Given that the evolutionary object is to adapt to, and exploit, predictable environmental cycles, it seems counter-productive to evolve a system that prevents any such adaptation. This analysis does, however, point out that stasis could be achieved—even in the rhythmic environment—with the proper phasing of opposing rhythms.

So, how is it possible to have multiple rhythms with different phases within a single organism? One possibility would be to incorporate different coupling times between the driving oscillation (say

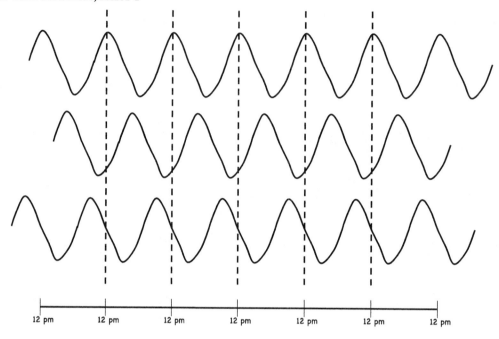

12 pm   12 pm   12 pm   12 pm   12 pm   12 pm   12 pm

Figure 3.7: Phase angle differences among rhythms.

X_{LD} in Figures 3.4 and 3.5) and the rhythmic systems or subsystems being driven. All that is necessary is to build in the proper delays between changes in X_{LD} and whatever changes in the biological subsystem in question and the necessary phase angles could be generated. Figure 3.10 displays this result. We can now begin to see a significant advantage to the new model—it provides a means of generating different phase relationships among various biological rhythms having the same frequency.

As a practical example of how this approach could be used, consider the effect of phase shifting the hepatic phosphorylase rhythm so that it is 180° out-of-phase with the rhythm in glycogen synthetase activity. For example, if these were twenty-four hour rhythms and both originally peaked at 12 noon, the hepatic enzyme activity would be phase shifted to peak at midnight and reach a trough at noon. Notice the effect on changes in glucose levels—when both rhythms are in phase, there are no changes, but when the two rhythms are 180° out-of-phase, there is a high amplitude rhythm, reflecting large changes in glucose concentrations (Figure 3.11). This may seem downright silly—why have the glycogen synthetase rhythm at all if all that is needed to create a glucose cycle is to change the activity of hepatic phosphorylase? The main reason is that the human body requires that glucose be stored as glycogen and without glycogen synthetase, that will not happen in liver. You must have both enzymes to move glucose back and forth between its storage form as glycogen and its useable form as glucose. Also, by changing the phase relationships (0° and 180° are not

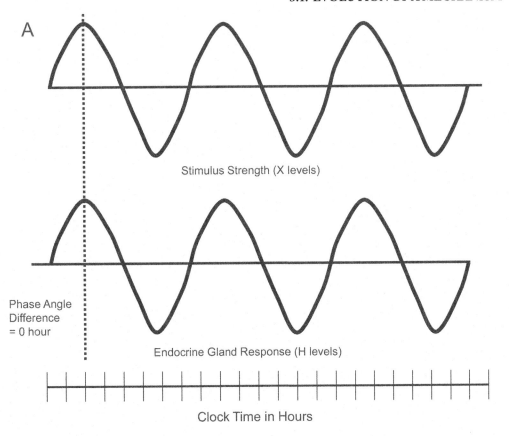

Figure 3.8: Phase angles between two rhythms. In (A) the phase angle is 0 hours. *(Continues.)*

the only possible phase angles), the amplitude of the glucose rhythms can be modified. I will defer discussion as to how this works to a later time. What this example does emphasize is that there is more to rhythms than just frequencies. Even if every environmentally driven biological rhythm had the proper frequency, there might still be the problem of ensuring the proper phase relationships. The new models illustrated in Figures 3.4 and 3.5 allow such phase relationships while ensuring a stable frequency.

3.1.5   REACTION VS. PREDICTION IN ECOSYSTEMS AND THE NEW MODEL

The new model presented in this chapter appears to have overcome two of the major limitations inherent in the initial model presented in Chapter 1. By linking (coupling) all daily rhythms to the light/dark cycle through LD sensors, the organism can be reasonably assured of stable frequencies in

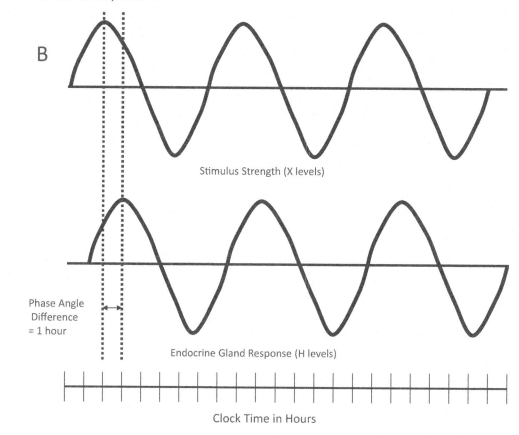

Figure 3.8:   *(Continued.)* Phase angles between two rhythms. In (B) the phase angle is 1 hour.

internal rhythms and can set the proper phase relationships between these various rhythms. What about the last problem, that of prediction vs. reaction in ecosystems? Can the new model(s) generate the proper external, as well as internal, temporal order?

Interestingly enough, the problem of prediction vs. reaction is actually another problem of establishing a proper phase relationship. In this case, the phase relationship involves three rhythms—the external rhythm driving the internal biological rhythm(s), the internal biological cycles, and the ecological rhythm to which the organism adapts. In our model, the driving environmental rhythm is the LD cycle, although this need not be the case under all circumstances. Let us examine both the on-time bird and the hare's seasonal fur change to determine if our new model can generate the adaptive predictability required for temporally stable ecosystems.

Figure 3.12 shows how a physiological rhythm might be used to predict the proper timing for worm hunting. This is the same figure as Figure 1.1 in Chapter 1 but with the addition of a

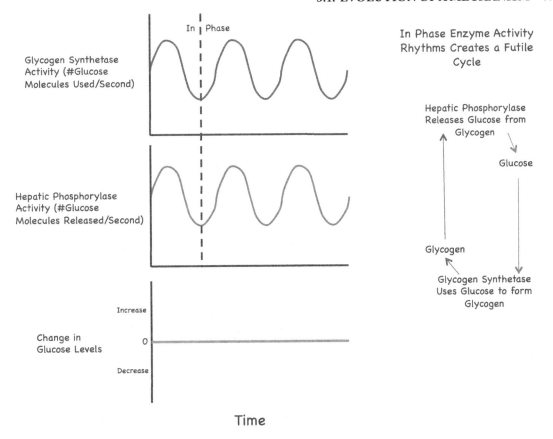

Figure 3.9: Effect of phase on the effectiveness of biochemical systems.

physiological rhythm shown as a red dotted line. This internal biological cycle is used to signal the approach of sunrise so that the bird in question can wake up and get to the feeding ground at the correct time. This internal biological rhythm is linked to the environmental light/dark cycle as shown in Figure 3.4, and this particular species of bird has evolved a mechanism of using the peak in this rhythm as a wake-up call. As long as the phase relationship between the internal physiological rhythm and the external environmental rhythm in worm availability is correct, the bird's behavior will be adaptive. The assumption underlying the model is that coupling this internal physiological rhythm with the stable geophysical cycle of day and night will allow the correct phase relationship (or phase angle) to be maintained.

You may have some personal experiences which appear to reflect this new model. If you work on a consistent schedule, you may find yourself waking up a few minutes before your alarm clock

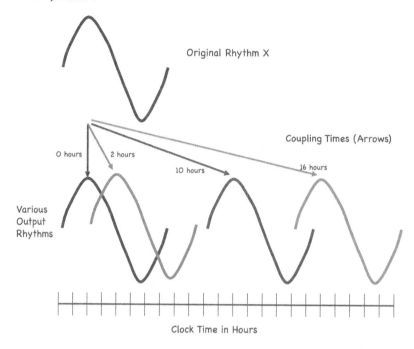

Figure 3.10: Changes in timing between stimulus and response creates different phase angles.

goes off. If effect, your body uses some kind of internal rhythm to anticipate the time the alarm would sound, analogous to our bird's awakening in time to get to the feeding grounds.

So far, so good. It appears that our new model can help with some aspects of external temporal order as well as internal temporal order. What about the seasonal rhythm of fur color change? Can the new model explain that as well?

For the model to work as proposed, the seasonal cycle must be sensed by the organism via a stable rhythmic change in environmental stimuli. For daily cycles, we used light and dark. What can be used for seasonal rhythms? There are number of environmental variables which change with the seasons—temperature, humidity, etc.—but few show the consistency necessary for the new model. One such predictable change would be that of the photoperiod or daylength.

How does daylength or photoperiod help in this situation? Table 3.1 lists the daylength for various geographic locations and dates throughout a calendar year. There is a predictable pattern of change for each location specified in the chart. In Seattle, Washington, for example, the photoperiod changes from a minimum day length of 9 hours and 12 minutes on January 1st to a maximum day length of 16 hours and 34 minutes on July 1st. Of course, the actual minimum and maximum day lengths occur on the winter and summer solstices, respectively, but Table 3.1 makes it clear that there

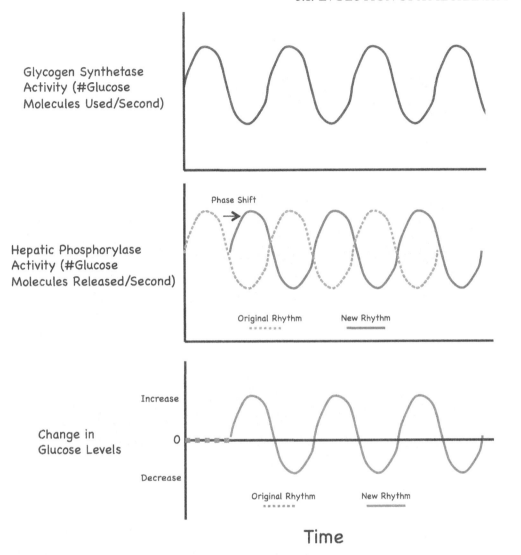

Figure 3.11: Effects of phase-shifting the hepatic phosphorylase activity rhythm on serum glucose levels.

is a rhythmic alteration in photoperiod which might be used as a stable indicator of season for an organism's internal annual rhythms. At least, theoretically.

Practically, however, we have a problem. Seasonal changes in photoperiod are latitude-dependent. Change an organism's location north or south and the photoperiod changes as well. If an organism is using the rhythm in photoperiod as a seasonal indicator, any significant travel north or south could throw off the timing signals to the internal rhythm and thus, throw the organism out

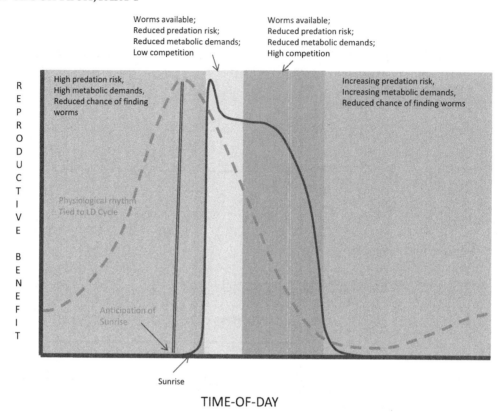

Figure 3.12: Application of a physiological rhythm to predict an important environmental change.

of synchronization with the external environment. Practically, then, photoperiod changes could be used as a timer only if the organism did not travel any significant distances north or south, i.e., had a limited north-south range. For organisms, such as plants, with little or no mobility, photoperiod changes should suffice as a seasonal indicator.

Even in organisms with a limited range, however, Table 3.1 indicates another problem. This time the problem stems from using the model to fit the organism into a daily pattern. The assumption underlying the new model was that the daily patterns of light and dark are stable and thus can serve as a kind of temporal "anchor" to which all rhythms could be associated. Table 3.1 and Figure 3.13 would seem to indicate that the assumption is not quite correct. We as humans may know that the Earth's rotation is consistently completed in a little less than 24 hours, but all organisms can sense are changes in their environment associated with that rotation. In the case of light and dark, the actual LD rhythms change from day to day. As Figure 3.13 demonstrates, day length not only changes from month to month but at different rates and to different extents at different latitudes.

Table 3.1: Daylengths from different locations throughout the calendar year. (From http://www.lpi.usra.edu/education/skytellers/seasons/activities/light.shtm.)

	Miami, FL	Brisbane, Australia	Nairobi, Kenya	Punta Arenas, Chile	Nome, Alaska	Singapore	Cape Town, South Africa	Seattle, WA	Vostok, Antarctica
Jan. 1	10h 34m	14h 31m	12h 12m	17h 32m	4h 53m	12h 04m	15h 03m	9h 12m	24h 00m
Feb. 1	11h 00m	13h 22m	12h 10m	15h 27m	7h 02m	12h 04m	13h 45m	10h 15m	24h 00m
March 1	12h 18m	13h 20m	12h 08m	13h 33m	10h 03m	2h 05m	13h 20m	11h 04m	19h 01m
April 1	12h 25m	12h 29m	12h 06m	11h 22m	13h 35m	12h 07m	12h 24m	13h 31m	14h 40m
May 1	13h 07m	11h 04m	12h 04m	10h 03m	17h 36m	12h 50m	11h 26m	15h 18m	0h 00m
June 1	14h 18m	11h 1m	12h 03m	8h 32m	21h 17m	12h 51m	10h 03m	15h 42m	0h 00m
July 1	14h 24m	11h 05m	12h 03m	8h 17m	22h 09m	12h 11m	10h 36m	15h 36m	0h 00m
Aug. 1	14h 00m	11h 29m	124 04m	9h 27m	18h 04m	12h 10m	11h 07m	15h 36m	0h 00m
Sept. 1	12h 40m	11h 32m	12h 05m	11h 22m	15h 16m	12h 08m	11h 23m	13h 22m	7h 14m
Oct. 1	12h 35m	12h 20m	12h 07m	12h 46m	11h 19m	12h 06m	12h 24m	11h 41m	15h 14m
Nov. 1	11h 11m	13h 49m	12h 09m	14h 53m	8h 36m	12h 04m	14h 08m	10h 39m	24h 00m
Dec. 1	11h 20m	14h 24m	12h 09m	16h 33m	5h 31m	12h 03m	14h 14m	9h 24m	24h 00m

In my home city of Philadelphia, for example, the sunrise occurred at 5:36am on July 1, 2008, and sunset occurred 14 hours and 57 minutes later. On December 1, 2008, the numbers were 7:04am and 9 hours and 32 minutes. How is an organism going to maintain external temporal stability in the face of such a changing environment?

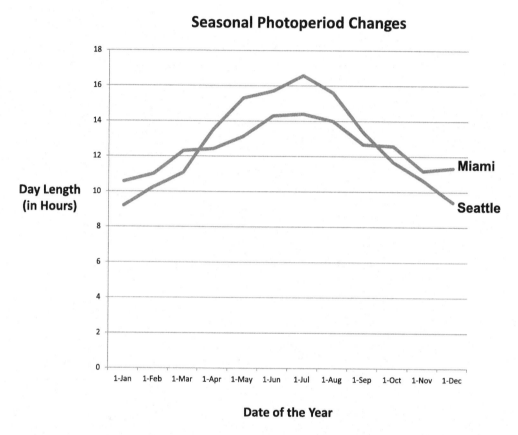

Figure 3.13: Seasonal photoperiod changes at two different locations.

Luckily, the day-to-day changes are not all that large. For example, in Miami the change during September and October is about a minute and a half per day. Even in places, such as Seattle, subjected to greater rates of change, the daily difference remains low at around 3 minutes per day during times of maximum change. These changes are probably not significant in terms of throwing off external temporal order for an organism located there. In addition, such changes are sufficiently gradual that the internal rhythms could adjust to these gradual LD changes while retaining the proper phase relationships.

Maintaining the proper phase relationship between internal and external rhythms is critical— it is how the on-time bird remains on time and how the hare prepares its winter fur for the winter snows. We see now that this relationship is not entirely fixed but must itself adjust to seasonal changes in photoperiod. For example, consider the situation of a nocturnal rodent who is active at night to avoid various dangers—such as predators and competitors—during the daylight hours. Suppose this creature's internal rhythm is set for wake-up 12 hours after sunrise. Let us further suppose this animal lives near Seattle, WA. A 12-hour phase angle between sunrise and activity onset makes sense for the fall and winter months of October through March but has the animal active during daylight hours in the spring and summer (April through September). In the latter case, the animal is exposed to dangers during those daylight hours that could be avoided by changing the phase angle from 12 hours to 17 hours. In fact, it could be argued that to gain the maximum benefit from temporal adjustment to the prevailing ecosystem, the phase angle should gradually adjust throughout the entire year. This means that any determination of the proper phase relationships for daily rhythms is dependent on seasonal cycles. For the majority of organisms, apparently, one timer is never enough!

3.2 CHAPTER REVIEW

In this chapter, the question of whether or not environmentally driven biological rhythms could be adequately coordinated to generate internal and external temporal order was examined. The approach using multiple environmental cycles, each driving an internal biological rhythm specific to the character of the oscillating environmental stimuli, did not prove able to generate the necessary temporal order. A new model was proposed, where certain select environmental cycles—such as day/night or seasonal photoperiod changes—are used to drive the entire internal rhythm system. By limiting the number of input rhythms, and by building in specific phase delays, internal temporal order could be achieved if the driving environmental cycle was sufficiently stable. However, the light/dark cycle associated with day and night is not—from an organism's point of view—entirely stable and thus changes in phase angle will be required to adjust to photoperiod changes throughout the year. These seasonal photoperiod changes make it difficult to achieve external temporal order with the proposed model.

Environmentally driven biological rhythms, even when the number of environmental drivers are limited, do not appear sufficient to generate reliable internal and external temporal order. An alternative approach—self-generating biological clocks—will be explored in the next chapter.

REFERENCES

[1] Angel, A. (1989). The effects of space travel on the nervous system. *Journal of the British Interplanetary Society* 42(7): 367–370. 66

[2] Barton, N.H., Briggs, D.E.G., Eisen, J.A., Goldstein, D.B. and Patel, N.H. (2007). *Evolution.* Cold Spring Harbor Laboratory Press: Cold Spring Harbor, New York, pp. 509–511. 65

[3] Gunn, C.K. (1932). Color and primeness in variable mammals. *The American Naturalist* 66(707): 546–559. DOI: 10.1086/280460 69

86   REFERENCES

[4] Hawkey, A. (2003). The physical price of a ticket into space. *Journal of the British Interplanetary Society* 56(5–6): 152–159. 66

[5] Hawkey, A. (2005). Physiological and biomechanical considerations for a human Mars mission. *Journal of the British Interplanetary Society* 58(3–4): 117–130. 66

[6] Holland, H.D. "Atmosphere, evolution of," in AccessScience@McGraw-Hill. `http://www.accessscience.com.ezproxy2.library.drexel.edu`, DOI10.1036/1097--8542.058900 65

[7] Payne, M.W., Williams, D.R. and Trudel, G. (2007). Space flight rehabilitation. *American Journal of Physical Medicine and Rehabilitation* 86(7): 583–591. DOI: 10.1097/PHM.0b013e31802b8d09 66

[8] Rowe, W.J. (1997). Interplanetary travel and permanent injury to normal heart. *Acta Astronautica* 40(10): 719–722. DOI: 10.1016/S0094-5765(97)00170-7 66

[9] Vandenburgh, H, Chromiak, J., Shansky, J., Del Tatto, M. and Lemaire, J. (1999). Space travel directly induces skeletal muscle atrophy. *The FASEB Journal* 13(9): 1031–1038. DOI: 10.1016/S0094-5765(97)00170-7 66

CHAPTER 4

Clock Tech II From External to Internal Timers

Overview

In this chapter, we will further analyze the model developed in Chapter 3 using the *light/dark cycle* as the sole driving environmental oscillation *generating daily biological rhythms*. Despite the improvements afforded by this approach over multiple environmental cycles driving multiple biological rhythms, *problems associated with maintaining both internal and external synchrony remain*. This will lead us to consider the *heart as a model oscillatory system*. The heart must solve the twin problems of maintaining a coherent and stable overall cycle while simultaneously being sufficiently flexible to meet the changing blood flow and metabolic demands of the organism. To accomplish these twin tasks, the heart evolved a centralized set of *pacemaker cells* that are capable of generating oscillations without any cycling environmental input, thus maintaining a stable rhythm. To understand how this evolved, we will investigate the *basic mechanism of action potentials* only to discover a new form of biological rhythm, one that could be initiated with constant, rather than oscillatory, environmental input. Further investigation will demonstrate that relatively minor changes in the basic parameters of action potentials can result in the *evolution of autorhythmic neurons*. These neurons could be adjusted by environmental cycles to create both a *frequency match* to the environmental cycle and a *stable phase angle* between the two oscillations. Insofar as these are the conditions necessary to create external synchrony, we will turn our attention to a *new model of biological rhythms* associated with geophysical cycles, *autorhythmic or endogenous pacemakers*. This new model of generating synchronized biological rhythms will show not only a method of *maintaining external synchrony*, but significant advantages over environmentally driven rhythms in *maintaining internal synchrony* as well.

The only reason for time is so that everything doesn't happen at once.

–Albert Einstein

One always has time enough, if one will apply it well.

–Johann Wolfgang von Goethe

4.1   THE COST AND BENEFITS OF ENVIRONMENTAL DRIVERS

In Chapter 3, we examined the problems of maintaining rhythmic coherence in the face of multiple environmental cycles. A model based upon multiple environmental cycles driving multiple physiological rhythms suffers from the tendency to generate internal desynchronization due to variations in the different environmental cycles. To compensate for this, the model was altered to allow for a limited number of environmental drivers, or in our specific case, only one—the light/dark cycle. Two variations in this model are possible—many LD sensors, each linked to a limited number of physiological systems or a few centralized LD sensors (or even one) linked to the entire organism. Each version has its own advantages and disadvantages. Multiple LD sensors create system redundancy such that the loss of a single sensor will not cause all internal rhythms to lose external synchronization. On the other hand, multiple sensors will inevitably result in slight variations in sensitivity and other characteristics, increasing the difficulty of maintaining internal synchronization among various physiological rhythms. A single or limited number of centralized LD sensors would not suffer from the latter problem insofar as all physiological rhythms will be linked to a limited number of rhythm drivers. Such a structure does create an increased risk of system-wide failure, however, since all physiological rhythms would be affected by damage or loss of the primary LD sensors. In both models, the system design relies completely on the light/dark or day/night cycle as the instigator generating and maintaining all internal physiological rhythms. The difference between the new models and the old one is that in the previous model, each physiological system responded directly to its own stimulating environmental cycle whereas in the new versions, the stimulating environmental cycle for all physiological rhythms is the alternation of light and darkness. In that sense, the new models are the biological equivalent of time estimation via a sundial.

The biological models do have one advantage over using a sundial. Sundials do not oscillate on their own, whereas our simulations indicate that physiological rhythms will persist for several cycles after the environmental cycle is removed. In an effort to determine just how persistent such rhythms might be, we altered our model to make it a bit closer to the physiological reality. We modified the environmental stimulus, X, to create two components, an external X reflecting the actual environmental cycle and an internal X which transmits the environmental cycle information to the rest of the system. In this way, we are a bit closer to the new models based upon the use of LD cycles. In addition, we modified the endocrine gland such that the levels of hormone, H, reflect a setpoint for H. This provides two setpoints, one for the internal X factor and another for the levels of hormone, H. Of course, there are hundreds of different kinds of physiological systems whose rhythmicity must be maintained and synchronized, so this model is only a primitive approximation of the real system. However, even primitive models can be revealing, and such was the case here.

Before examining the effects of removing a rhythmic environmental stimulus, we needed to confirm our previous observations that an environmental rhythm could still drive the system into a cycling state. Figure 4.1 shows that the changes in the model described above did not alter that fundamental result. In the figure, a constant level of stimulus was applied to the system from Time=0 to Time=50 (arbitrary units). However, at Time=50, the constant level of stimulus was replaced with a rhythmic stimulus. Oscillations are produced in all system components at the same time (Time=50). The stimulus plotted here is the internal stimulus, X_{in}, since that rhythm is what drives the rest of the components. It is assumed that the external stimulus, X_{ext}, directly stimulates X_{in} to force the oscillation in X_{in}.

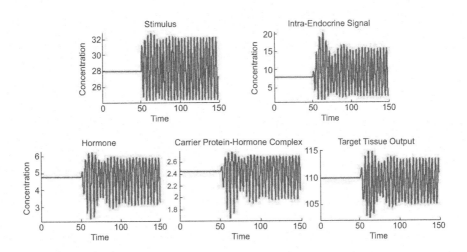

Figure 4.1: System response to a rhythmic stimulus.

What, then, occurs when the rhythmic stimulus ceases? Does the system continue to oscillate and for how long? This question is answered in Figure 4.2. In the latter figure, a rhythmic stimulus was applied to the system from Time 0 to 50. However, at Time=50, the rhythmic stimulus was replaced with a constant level of stimulus. For moments after the rhythmic stimulus is removed, the four system components shown below (intra-endocrine signaling molecule, hormone, carrier-protein—hormone complex, and the target tissue output) continue to oscillate before coming to a steady state. The stimulus plotted here is the external stimulus, X_{ext}.

From the plots, it is evident that the system components oscillate for several cycles, but with a rapidly decreasing amplitude. This approach may be a bit better than a sundial, but not by very much. To rely on these damped oscillations to maintain internal and external synchrony when an organism is isolated from a light/dark cycle would be a risky proposition at best.

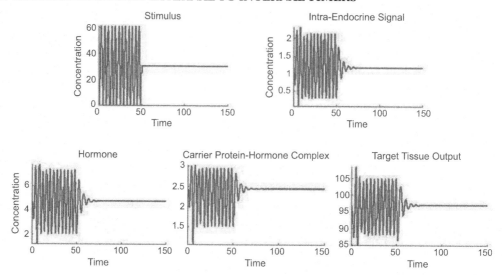

Figure 4.2: System showing a damped rhythm after rhythmic stimulus is removed at time T=50.

Still, when would an organism be isolated from the changes in light intensity associated with day and night? Would this really pose any significant limitation or risk to organisms?

This turns out to be a more difficult question to answer than one might expect. First, there are situations in which the behavior of an organism might isolate it from the light/dark cycle—being deep underwater, in a burrow or a cave, for example, or even under significant forest cover. Second, there is the matter of how cycling light intensity information is transduced into a rhythm in X_{in}. The actual environmental light intensity reaching an organism will vary depending on various environmental conditions. A significant cloud cover could easily change the amplitude of the light/dark cycle with unforeseeable consequences.

In Figure 4.3, the latter problem is graphically displayed. The graphs plot time-of-day vs. light intensity for three consecutive days. Day 1 has a light cloud cover and moderate light intensity, Day 2 is heavily overcast, and Day 3 is bright and clear. As can be seen from the graphs, if dawn is recognized by the time at which a certain light intensity occurs (indicated by the threshold line) then dawn occurs at different times on Days 2 and 3 compared to Day 1 as indicated by the vertical blue line. Not only would this potentially cause phasing problems, as each dawn may be different relative to the important environmental events the organism wishes to anticipate, the perceived frequency of the environmental cycle varies as well. For the sake of simplicity, assume that the variations in sunrise indicated by the vertical blue line are an hour long. The actual perceived cycle length (from perceived dawn to perceived dawn) from Day 1 to Day 2 is then 25 hours long and the time from Day 2 to Day 3 is 22 hours long. The constantly changing frequency could disrupt the organism's attempt to maintain internal temporal order.

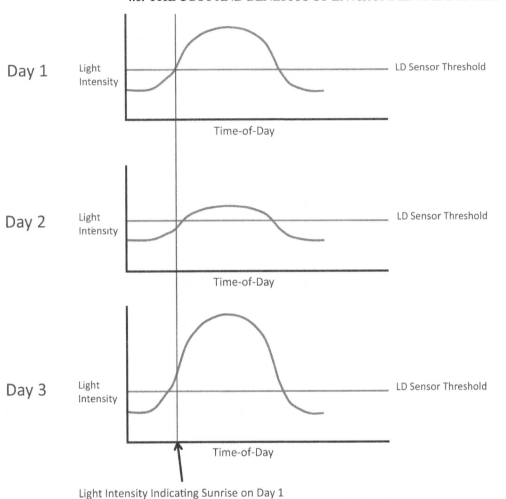

Figure 4.3: Effects of light intensity variations on transduction of day/night cycles onto physiological rhythms.

In addition to these issues, there is the problem of seasonal changes in both photoperiod and light intensity. The photoperiod problem was discussed in Chapter 3, but light intensity varies with seasons as well as light duration, compounding the problem of maintaining external synchronization.

To review, creating a more manageable system by which all physiological rhythms are linked to the day/night cycle reduces the potential for internal desynchronization but does not eliminate it. Variations in light intensity rhythms could create instabilities in rhythm coherence when directly linked to internal cycles. In addition, being completely dependent upon day/night variations seems to make the system almost too sensitive—slight variations in the physical environment have potential impacts on physiological rhythmicity out of proportion with the original variation. Finally, seasonal changes in photoperiod and light intensities create difficulties in maintaining the proper phase relationship between internal physiological and behavioral rhythms and external environmental events. In effect, to know what time it is, an organism would have to know what season it was—which begs the question of how the organism anticipated seasonal changes in the first place.

4.2   A NEW DAWN—INTERNALIZING THE TIMERS

The problem organisms face on Earth is they have a set of conflicting temporal requirements. On the one hand, as complex devices, they must maintain a stable internal temporal organization, similar to an automobile or computer. On the other hand, unlike cars or computers, organisms must synchronize their internal rhythms with external environmental events. The variability of most environmental rhythms make them unsuitable to drive internal cycles—this variability would translate into variability of internal cycles, resulting in a loss of internal temporal order. However, even the most stable external cycles, such as the day/night cycle, remain variable *as perceived by organisms*. Thus, using only one stable environmental cycle as a driver of internal rhythms reduces but does not eliminate the potential for internal desynchronization. The question really boils down to this—how sensitive should the internal rhythm system be to the environmental cycles to which it must synchronize? If the system is too *sensitive* to slight variations, normal changes in daily light intensity will be magnified into rapidly altering rhythmic signals producing internal synchronization. If the system is too *insensitive*, it will be unable to respond to important changes in environmental lighting, such as seasonal alterations in photoperiod, and thus become externally desynchronized.

The evidence presented in Chapter 2 demonstrates that organisms have evolved solutions to these conflicting requirements. How can a rhythmic system evolve which is simultaneously stable and flexible with both internal and external synchrony? Perhaps we have been approaching the problem from the wrong direction. In Chapter 3, we took an external point of view, focusing the effects of environmental cycles on biological systems. However, our models indicate that making any component of the system oscillate results in all components of the system becoming rhythmic. Figure 4.4 displays this effect with the revised model. Thus, we could analyze the system from an internal point of view and examine the results of having the oscillations *endogenously*, rather than *exogenously*, generated.

Do we have any simple physiological systems that we can use as examples to clarify the differences between these two approaches? Although simple may not be an entirely appropriate term, cardiac and respiratory rhythms can be used to illustrate these differences. At first glance, this may seem a bit strange. How can two internal biological cycles be used to illustrate the differences between endogenous and exogenous rhythms? Aren't both rhythms endogenous?

The answer is that—as often is the case—it depends. From the point of view of the entire organism, it is true that both rhythms are endogenous. From the point of view of the muscle tissue and organs that cycle, however, the situation is quite different. Cardiac rhythms are endogenously

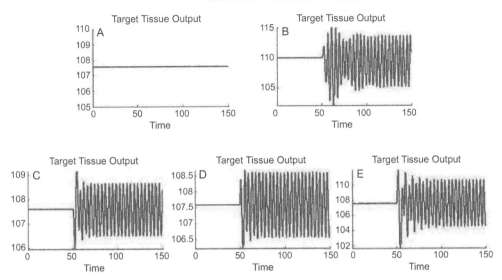

Figure 4.4: Target tissue output of systems where only one component displays an endogenous rhythm. In plot A, there are no system components that are displaying an endogenous rhythm and no rhythm is seen in the output. In plot B, the stimulus is rhythmic and the output of the system is also rhythmic. In plot C, the enzyme in the endocrine gland is endogenously rhythmic and produces oscillations in the system's output. In plot D, the total concentration of carrier protein is endogenously rhythmic and produces oscillations in the system's output. In plot E, the enzyme in the liver which degrades proteins in the blood is endogenously rhythmic and also produces oscillations in the system's output.

generated by cardiac muscle fibers. This can be demonstrated by the observation that hearts removed from the body will continue to beat. The rhythm is an inherent characteristic of the heart itself. On the other hand, if the lungs and diaphragm were to be removed, breathing would cease. Stimulation from motor neurons in the spinal cord is necessary to drive the respiratory rhythms—there is no inherent cycling component within the lungs or respiratory musculature. In the case of respiration, a rhythm is imposed upon the lungs and associated muscles by the central nervous system. That makes the rhythm exogenously generated from the point of view of the lungs and respiratory muscles.

Considering the heart first, the timing mechanism for that organ has evolved to be located in specialized sections—the nodes—which serve as both timer and link to outside stimuli (outside of the heart, that is). The heart has several *autorhythmic* components—the sinoatrial (SA) node, the atrioventricular (AV) node, and the various elements of the conduction system. Normally, the dominant frequency of the SA node forces a coherent rhythm over the entire heart. This organization prevents extraneous stimuli from desynchronizing the various heart components while providing for some flexibility in frequency and amplitude of the biological cycle. In addition, such an approach provides for centralized control of the overall rhythmic system. What if this was not the case? What if each oscillating component had its own, independent input capable of generating significantly different phases in those components? Different sections of the heart would depolarize and contract at different times with no guarantee that any given contraction would be coordinated with any other. Clearly, the heart could not continue to function properly under such conditions. By having a limited number of nodes, it is possible to regulate heart rate by stimulating or inhibiting (increasing or decreasing the frequency of) the nodes rather than connecting to every individual fiber. The endogenous character of the rhythm means that external influences are limited mainly to adapting the rhythm to new circumstances. External factors are not needed to generate a rhythm and thus the cardiac cycle is protected against the impact of any interruptions in such external stimulation. This is the most effective approach if the timing system is going combine extreme stability with some level of adaptability.

The situation with the respiratory system is very different. There are no autorhythmic components in the lungs or associated respiratory muscles. The respiratory rhythm is imposed on these tissues through actions involving the central nervous system. If the heart rhythm model is so successful, why was it not used for respiration as well? Consider the different contexts within which the two organ systems operate. The heart rhythm drives blood throughout the body and cannot be interrupted for any significant period of time without serious consequences, especially to the brain. In addition, the heart has a single purpose—pumping blood—and was able to become specialized for that purpose. The respiratory system is functionally more diverse. Airflow through the system is used for communication as well as respiration and this requires considerable voluntary control. The respiratory system also connects with the digestive system (which is why you cannot eat or drink and breathe at the same time).

An autorhythmic respiratory cycle constructed similar to the heart would be very difficult to coordinate with voluntary consumption. In such a system, the lungs would operate automatically and with their own rhythm, unable to be stopped, and thus each act of swallowing would require a person to precisely time the event in order to swallow between breaths. Choking would be far more likely with such an organization. Finally, breathing is a bit less time critical than pumping blood and can be interrupted for somewhat longer periods. Holding your breath is one thing, stopping your heart is quite another. This flexibility comes with a price, of course, because the system is wholly dependent on input from the central nervous system. If that input is interrupted for any reason, the respiratory rhythm ceases. This is why individuals suffering from spinal injuries associated with a broken neck require a respirator and not cardiac stimulation.

Let's now zoom out from our specific examples and return to the level of the whole organism. The respiratory situation is analogous to what we have discussed so far—environmental rhythms driving physiological rhythms. The heart, on the other hand, is a new kind of approach with an internal oscillator serving as a temporal interface between the external environment and internal physiological system. Which would be the better approach for organisms to evolve in order to resolve the twin issues of external and internal synchronization?

Well, we have already seen that environmental drivers—even when reduced to only one, the LD cycle—generate some potential problems in maintaining the required temporal synchronizations. The heart model does not seem to suffer from these problems, at least internally. Can this model be adapted for the more complex problem of maintaining internal temporal order while simultaneously adjusting to important geophysical cycles? To answer this question, we need to further investigate the nature of cardiac rhythmicity.

4.3   ENDOGENOUS RHYTHMS AND BIOLOGICAL CLOCKS

How do the autorhythmic cardiac myocytes (muscle cells) generate an endogenous rhythm? To understand this, we need to review—at a simple level—the nature of electrical potentials and how these potentials affect excitable tissue—nerve and muscle cells. This discussion is based upon Gary Matthew's [1] excellent presentation. Anyone familiar with these ideas can profitably skip over this section.

4.3.1   EXCITABLE TISSUE

All animal cells have an electrical potential distributed across their cell membranes. This potential comes from a very slight imbalance of charged atoms and molecules across theses membranes. To create a potential difference of 100 mV (millivolts)—somewhat greater than the usual potential— only requires an imbalance of one charge in a billion. However, as tiny as this slight imbalance seems, it represents a real physiologically relevant difference that results in an interesting balance of forces across the membrane.

Furthermore, the actual ions and molecules contributing to this electrical potential are different inside and outside of the cell. On the inside, the potential comes mostly from positively charged potassium ions (K^+) and a host of different organic molecules (proteins, carbohydrates, lipids, nucleic acids, etc.) collectively referred to as negatively charged organic anions (A^-). The external potential comes primarily from positively charged sodiums (Na^+) and negatively charged chlorides (Cl^-). Figure 4.5 displays some typical values and distributions. The distributions create a situation where the cell is both osmotically balanced (i.e., water is not prone to move either into or out of the cell) and electrically balanced (the number of charges inside and outside of the cell are balanced). How does this accord with my claim above that an electrical potential exists across all such membranes? Doesn't Figure 4.5 contradict that claim? The answer lies in the size of the electrical potential. As indicated above, only one charge imbalance out of a billion is required to create a 100 mV potential. That kind of difference is too small to be reflected in the millimolar (mM) concentrations shown in Figure 4.5.

The cell expends a great deal of energy maintaining this distribution. Na^+ and K^+ tend to leak down their concentration gradients and must constantly be moved back across the membrane. This is carried out by the actions of an energy-requiring protein called the sodium-potassium pump.

Now, suppose the cell starts out as electrically neutral on a microscale as well as on the macroscale. In other words, the actual number of positive and negative charges were equal on both

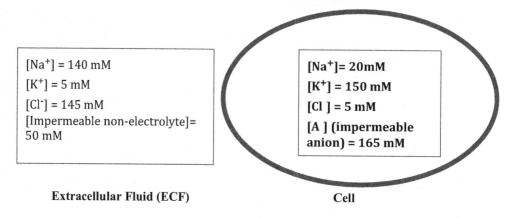

[Na⁺] = 140 mM

[K⁺] = 5 mM

[Cl⁻] = 145 mM

[Impermeable non-electrolyte]= 50 mM

[Na⁺]= 20mM

[K⁺] = 150 mM

[Cl] = 5 mM

[A] (impermeable anion) = 165 mM

Extracellular Fluid (ECF)   **Cell**

Figure 4.5: Typical distributions of charged atoms and molecules between animal cell and the external fluid. (Redrawn from Michael and Rovick[2].).

sides. If K^+ began to diffuse down its concentration gradient (from inside the cell to the ECF), each time a K^+ left, there would be a unit decrease in the number of positive charges inside the cell. In effect, the inside of the cell would appear negatively charged relative to the outside. This creates an electrical potential, albeit a small one, across the membrane. The more K^+ diffuses, the greater the electrical potential becomes and the less force is exerted by the concentration gradient. Eventually, the two forces exactly balance, and diffusion ceases as equilibrium develops.

This situation is diagrammed in Figure 4.6. At the start, the only force (indicated by the direction and thickness of the arrows) is associated with the differences in K^+ concentrations and leads to diffusion. As diffusion proceeds, however, an electrical potential forms creating an electrostatic force opposing the diffusion force. Eventually, the two forces balance, as indicated by the arrows, becoming the same thickness. At that point, the two forces (arrows) are equal and opposite, and the system goes into equilibrium.

There are many ways to analyze this, but a particularly useful method is to determine the electrical potential at which equilibrium occurs. The equation is called the Nernst equation:

$$E_K = (RT/zF) * \ln\left\{[K^+]_{out}/[K^+]_{in}\right\} \tag{4.1}$$

Where E_K is the equilibrium potential for potassium, R is the gas constant (in our case, we will use 1.987 calories/mole-degree), T is the absolute temperature (for humans, we will use 98.6°F which is 310°K), z is the charge on the atom or molecule (+1 for potassium), and F is Faraday's constant (we will use 23062 calories/volt-mole). A mole of something is 6.023 x 10²³ somethings, such as 6.023 x 10²³ potassium ions.

If we go ahead and plug the values from Figure 4.5 into the equation, we find that E_K equals about −91 mV. We can also select another ion, such as sodium (Na^+) and plug those values in and

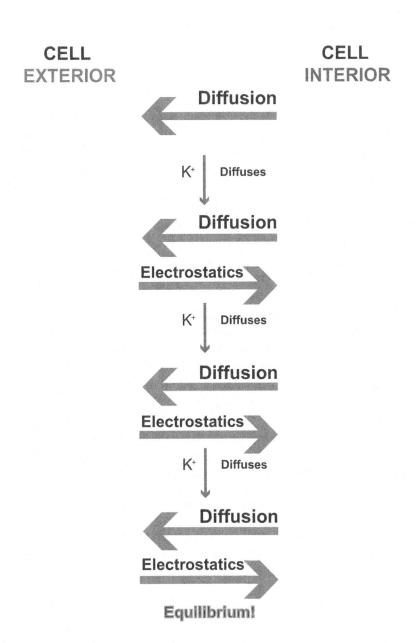

Figure 4.6: Effects of potassium (K^+) diffusion on cell membrane and potential.

calculate the equilibrium potential for Na^+. If we do that, the E_{Na} turns out to be about $+52$ mV. So, what is the actual membrane potential, -91 mV or $+52$ mV?

Well, if we were to actually measure the membrane potential in such a cell, we would find it to be slightly more positive than the E_K, somewhere in the range of -85 mV to -90 mV, depending on the specific cell. Clearly, the membrane potential is not at equilibrium as calculated for either potassium or sodium. How is this possible?

If we ignore chloride (Cl^-) for the time being and concentrate solely on sodium and potassium, what do these values indicate? It seems that the measured potential is close to that calculated as the equilibrium (or Nernst) potential for K^+ but slightly positive, in the direction of E_{Na}. At first glance, one is tempted to conclude that K^+ is almost passively diffusing through the membrane but Na^+ is barely leaking through, contributing in a very small way to making the membrane potential a bit more positive. However, ions do not really passively diffuse through biological membranes at all—the only way such charged molecules can get through the membrane is through protein channels specifically evolved to allow them to cross. Restating our original idea with this new condition in mind, we estimate that there are many potassium channels, allowing considerable potassium diffusion but only a few sodium channels, allowing far less significant sodium diffusion.

Okay, then, there is diffusion in both K^+ and Na^+. Interestingly, if one were to record the membrane potential of such a cell as displayed in Figure 4.5 for a long period of time, the membrane potential (V_M) would not change. It would remain steady and stable. If ions are moving across the membrane—and an unbalanced distribution of charges created the potential in the first place— why does V_M not change, reflecting this movement of charged atoms? One possibility is that the movement of K^+ and Na^+ is simultaneously equal in magnitude and opposite in direction. If for every K^+ diffusing out of the cell, there was a Na^+ moving in, then the distribution of charges inside and outside of the cell would remain the same, and thus V_M would remain the same as well.

We can actually develop an equation reflecting this possibility by recalling that any movement of charged particles—be they ions or electrons—creates a current. Currents are calculated using Ohm's Law, which is given by the following formula:

$$I(current) = V(voltage)/R(Resistance) \qquad (4.2)$$

So far, so good. Let's say we're not quite certain what the resistance of a biological membrane is, so I am going to modify the formula a little bit and substitute conductance for resistance. Conductance is just the inverse of resistance, and the formula becomes:

$$I(current) = V(voltage)/R(Resistance) = V(voltage) * g(conductance) \qquad (4.3)$$

Our idea is that the currents created by sodium and potassium will be equal and opposite. This is given by the following equation:

$$I_K = -I_{Na} \qquad (4.4)$$
$$= g_K(V_M - E_K) = -g_{Na}(V_M - E_{Na}) \qquad (4.5)$$

You may be a bit concerned about the value in the bold parentheses. The g values make sense—these represent the conductance of potassium and sodium and must reflect, at least in part, how many open channels exist for each ion. However, why is the voltage the *difference* between the membrane potential V_M and the equilibrium potentials? Why not just use V_M?

Go back and take a look at Figure 4.6. The arrows for diffusion and electrostatics represent two forces being applied to the potassium ions. One force—diffusion—pushes K^+ out of the cell while the other force—electrostatics—pushes K^+ into the cell. The electrostatic force is the result of the membrane potential. Now, at equilibrium as shown at the bottom of the figure, there is no net force on the potassium—the diffusion and electrostatics arrows are exactly the same size and aimed in opposite directions. They are equal and opposite and thus there will be no net movement of K^+.

However, we are suggesting that a K^+ current exists—that there is a net movement of potassium. To do this, there must be an electrostatic force on K^+ greater than zero, greater than the electrostatic force associated with E_K when there is no net force on potassium. It is the difference between the membrane potential, V_M, and the equilibrium potential, E_K, that generates this force, and that is what we must measure to determine the potassium current. The same argument holds for the sodium current.

Equation 4.5 can be rearranged to use it to determine the membrane potential given the Nernst potentials for sodium and potassium and their relative conductances. The result is provided below:

$$V_M = \left\{ g_K/(g_K + g_{Na}) \right\} * E_K + \left\{ g_{Na}/(g_K + g_{Na}) \right\} * E_{Na} \tag{4.6}$$

This equation has a rather interesting consequence. If you were to set the conductance of sodium (g_{Na}) to zero, the membrane potential, V_M, would become equal to E_K. On the other hand, if g_K were set to 0, V_M would become E_{Na}. Theoretically, at least, all that would be needed to set the membrane potential of any cell from any point ranging from the equilibrium potential of potassium (-91 mV in our example) to the equilibrium potential of sodium ($+52$ mV using the data from Figure 4.5) is the ability to alter the sodium and potassium conductances (g_K and g_{Na}).

In fact, this is far more than merely theoretical. Channel proteins for sodium and potassium can be "gated"—open or closed to ion movements, depending on the conditions of the cell and its surroundings. Two particular kinds of such channels—voltage-gated sodium and potassium channels—are found on excitable cells, such as neurons and muscle cells. These gated channels alter the conductances of sodium and potassium, creating an electrotonic signal.

Here is how this works. The voltage-gated (or potential-sensitive) sodium channels are proteins that span the plasma membranes of excitable cells. These are complex proteins with two types of "gates"—an *m* gate and an *h* gate. If either gate is in the closed position, Na^+ will be unable to traverse the membrane through that protein. In effect, this reduces g_{Na} to 0 through that channel. Now, these proteins are dynamic and can open and close. Under resting condition, the *m* gates are closed and the *h* gates are open, a condition which prevents any sodium from passing through the channels. If the V_M is raised—made more positive—the *m* gates rapidly begin to open. This increases the sodium current (I_{Na}) through an increase in g_{Na} (see Equation 4.5). Since the I_{Na} is directed inward, Na^+ will move into the cell, raising V_M even more. This, in turn, causes more ***m*** gates to open, increasing I_{Na} and so on. The result is a positive feedback system which causes the membrane potential to rapidly ascend toward E_{Na}.

Other things are also occurring. The ***h*** gates on the sodium channels begin to close, reducing g_{Na}. At the same time, the ***n*** gates on voltage-gated *potassium* channels open, increasing the outward directed potassium current (I_K) through an increase in g_K. This current is made even greater by the now positive V_M, which contributes through the (V_M—E_K) part of the formula for I_K. Thus, both the distance of the new membrane potential from the E_K and the increase in g_K lead to a significant potassium current. The I_K is directed outward, carrying K^+ out of the cell and reducing V_M back into negative territory. In fact, for a brief time, the cell closely approaches E_K as g_{Na} approaches zero. The membrane potential often undershoots the original V_M for a time before slowly rising back to the original level. The original potential, V_M, is called the *resting potential* when discussing excitable cells. The change in potential generated through the changes in g_K and g_{Na} mediated by the voltage-gated channels is called an *action potential*, *spike*, or *impulse*. Such an action potential is diagrammed in Figure 4.7.

There are a couple of new features in the diagram which we have yet to discuss. First, there is the potential level labeled *threshold* in the diagram. What is a threshold? Threshold is the point at which the sodium current overwhelms the potassium current, making an action potential essentially inevitable. Nice definition, but what does this actually mean? If you go back to Equation 4.5, you will see the formulas for the two competing currents. The two parameters which determine the magnitude of these currents are the conductances, ***g***, and the potential force **(V_M–E_{ion})**. When a neuron or muscle cell is stimulated, the result is that a certain number of sodium channels will open. This increases g_{Na} and thus the Na^+ current, I_{Na}. Since the Na^+ current is directed into the cell, the result of the current will be a rise in V_M. So far, this is just a review of the sodium part of the action potential. However, I skipped over something when I described the positive feedback system. There is still a potassium current, I_K, and it will also be affected by the change in V_M. Even if we assume that g_K remains fairly stable initially, the increase in V_M will increase the magnitude of (V_M–E_K) and thus the magnitude of the potassium current. For example, let's say the original V_M in our example was −86 mV, while the E_K was −91 mV. The magnitude of (V_M–E_K) = 5 mV. However, with an increase in sodium current due to an increase in g_{Na}, V_M becomes more positive. What happens if V_M reaches −80 mV? The magnitude of (V_M–E_K) is now 11 mV, more than twice the previous level. This means an increase in the potassium current. Since the K^+ current is directed out of the cell, K^+ will diffuse into the ECF, leaving the cell with a greater negative charge. In effect, I_K counter-balances I_{Na}—as the sodium current increases, so does the potassium current. This effect prevents initiation of the sodium positive feedback system until the increase in g_{Na} becomes so great that the increase in I_K cannot compensate. That point is the threshold for that specific cell.

Thresholds have some significant effects on the cell. First, they introduce a time-delay, however brief, between stimulation and response. Second, thresholds provide an adaptive intensity filter for neurons and muscle cells. Not every stimulus will be strong enough to push the cell above its

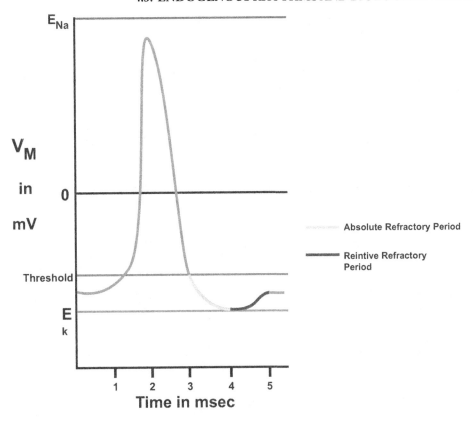

Figure 4.7: Diagram of an action potential.

threshold value. In fact, thresholds allows for each target cell to integrate all incoming stimuli to determine if the total summed stimulation reaches threshold or not. By having different cells evolve with different threshold points, a considerable amount of complexity can (and is) introduced into networks of excitable tissue.

The threshold is not the only new feature introduced in Figure 4.7. At the end of the potential changes indicating the action potential (blue line), there are two colored sections, labeled the *absolute* (yellow line) and *relative refractory* (purple line) periods. During the absolute refractory period, the sodium channels are no longer responsive to potential changes—thus, it is practically impossible to change the g_{Na}, which approaches zero. This is why the potential so closely approaches E_K— for intents and purposes, only the I_K is left to determine the cellular potential. As some sodium channels become responsive, it becomes possible to regenerate an action potential but only with a greater change in V_M. This is equivalent to raising the threshold—an action potential might be generated but only with greater stimulation. The latter time frame is called the relative refractory period.

Combining the effects of the threshold and refractory periods with the mechanisms leading to the generation of action potentials leads to a rather startling observation—excitable cells are natural oscillators! When stimulated at a constant level, excitable tissue generate action potentials with a frequency characteristic of the specific cell. This is displayed graphically in Figure 4.8.

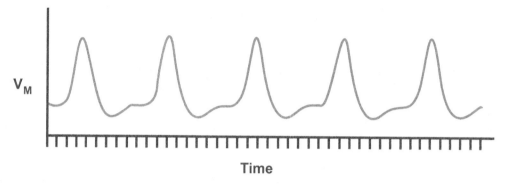

Figure 4.8: An oscillating sequence of action potentials in response to constant stimulation.

Thus, the fundamental structure of the nervous system is built from components that are, by their very nature, oscillatory. The brain and spinal cord—indeed, the entire nervous system—is made up of vast collection of neural oscillators.

4.4   A BRIEF REVIEW

At this point, we have really discussed two kinds of potential biological oscillators. Initially, we began by looking at physiological feedback systems and discovered that they all have the property that oscillations will develop in the system if one of the components begins to cycle. It does not matter what component initiates the process, the result will be the same—the system will oscillate. The component that we discussed the most, however, was that of the external environment which is dominated by several high-amplitude geophysical cycles. These cycles inevitably generate oscillations within the organisms exposed to them, resulting in multiple biological rhythms.

The other kind of oscillator is illustrated by the discussion of neuron function. In this case, the environment need not cycle to result in oscillation within the biological system. A constant level of stimulation into a neuron is sufficient to generate a rhythmic output. These cycles are not necessarily tied to any geophysical rhythms in the external environment.

We have also discussed two levels of adaptation that can be inferred from the existence of biological rhythms. One level of adaptation occurs when biological rhythms are associated with geophysical cycles. This synchrony allows organisms to prepare for predictable changes in their environment, providing these organisms with an adaptive advantage over species which merely react to change. The other adaptation reflects the inherent necessity of temporal order in any goal-directed complex device, whether a cell, a person, an automobile, or a computer. In order to optimize efficiency and effectiveness, complex devices must execute actions within a specified temporal sequence. This internal synchrony allows for complex devices to perform functions with maximum efficiency.

To review then, there are two vital functions for biological rhythms, internal synchrony and external synchrony. In addition, there are two primary mechanisms—the use of environmental cycles as drivers and systems that oscillate in response to a constant input. We will now turn our attention to a third potential mechanism—autorhythmic, endogenous oscillators.

4.5   FROM RHYTHMIC TO AUTORHYTHMIC: CREATING ENDOGENOUS OSCILLATORS

As we have seen through our previous discussions, the problem of creating both internal and external synchrony is not trivial. The use of multiple environmental drivers leads to multiple internal rhythms and the risk of internal desynchrony. Limiting the number of environmental drivers helps somewhat, but the risk of internal desynchronization remains a potential problem. In addition, the use of environmental drivers as the sole mechanism of creating external synchrony results in a signal sensitivity problem—if the organism is *too sensitive* to environmental signals, it becomes vulnerable to slight environmental changes generating significant alterations in biological rhythms, again raising the risk of *internal desynchronization*. If the organism is *not sensitive enough* to environmental changes, the result will be *external desynchronization*.

In turning to the heart, we were looking for a model of biological rhythmicity where internal synchrony could be maintained while still remaining sensitive to external factors—in this case, the requirement that cardiac output adjust to the needs of the body under various conditions. It is not a perfect model in that heart rate is not synchronizing to an environment cycle (as far as we know, there are no 60–70 cycle/minute geophysical cycles to which the heart normally synchronizes), but heart rate—cycle frequency—is adjustable. If you think about it, synchronizing an internal biological rhythm with an external geophysical cycle requires that the frequency of the two cycles match. Since adjusting the frequency of the Earth's rotation is pretty much out of the question, the only alternative is to adjust the frequency of the biological rhythm. Thus, using the heart as a model system is not as far-fetched as it might have at first appeared to be.

The evolution of biological oscillators has gone a step further in the heart than in the two rhythm-generating systems already discussed. There is no external cycle driving cardiac rhythms and the rhythms are not simply the output of a constant input stimulus. The latter is easy to demonstrate by the observation that a heart removed from the body will continue to beat despite no longer being associated with any physiological stimulus. How is this accomplished?

Muscle cells—like neurons—are excitable tissue and thus can develop action potentials. Certain cardiac myocytes (muscle cells) are autorhythmic, spontaneously depolarizing with specific time constants. How could such cells evolve?

A reexamination of the data from our discussion of neurons and action potentials suggests that we might be asking the wrong question. Perhaps it is not so mysterious that neurons—and excitable tissue in general—can be autorhythmic. The real question may be: How could organisms evolve to *prevent* such cells from cycling?

Consider the situation previously described. To maintain a resting potential requires a delicate balance of currents, a balance that must be actively maintained since the resting potential is not at equilibrium. To maintain this potential, the cell relies on the action of sodium and potassium channels and the activity of the sodium/potassium pump. Considering that generating a potential of 100 mV in a cell with a 100 μM diameter requires only a transfer of one ion per billion across the cell membrane [2] this balance would seem highly sensitive to any minute changes in ionic currents. For example, after an action potential, if fewer potassium channels opened or more sodium channels reopened than are required to maintain the resting potential, the cell's potential would drift upward and perhaps cross threshold, reinitiating an action potential. In fact, it would take considerable engineering to prevent this from happening.

Let us consider the situation from an evolutionary perspective. The ancestral cells from today's modern excitable tissue must have been less precise and controlled than their modern counterparts. In fact, the most primitive excitable cells must have been rather noisy since it would have taken considerable time and selection to develop a method of both generating action potentials and reliably returning to, and maintaining, a resting potential. It would be perhaps a bit too much of an evolutionary leap to move directly from a noisy, unreliable signal to a stable and reliable one. One potential transitional stage would be a cycling one, in which the noise level was controlled to the point of producing a predictable rhythmic pattern. The physiological issue would then be a method of controlling rhythmicity, rather than generating it.

Irrespective of the actual evolutionary pathway, the nature of action and resting potentials provides a simple conceptual mechanism for generating autorhythmicity. By changing the levels of one or both of the potassium and sodium currents such that threshold is reliably reached a certain time after each action potential, the cells become autorhythmic (Figures 4.9 and 4.11). In fact, such a mechanism also suggests a means of altering the timing of these cells to generate different frequencies. If we assume that the rate of potential rise from trough of the action potential is fairly stable, all that is necessary to change the frequency of autorhythmic excitable tissue is to alter the starting potential from which the threshold will eventually be reached. Lower the starting point, and it takes longer to reach threshold. Raise the starting point, and the threshold is reached earlier. This is displayed in Figure 4.10. In that figure, the original blue line represents the inherent frequency of the neuron oscillator, which is 1 spike or action potential every 4 msec. When the trough of the action potential is lowered, the frequency becomes 1 spike every 6 msec (purple line), while raising the trough (red line) gives a frequency of 1 spike every 2.5 msec. A mathematical simulation displaying the same result is provided in Figure 4.12, where a decrease in potassium conductance results in a increase in the firing on an autorhythmic neuron modeled in Figure 4.11.

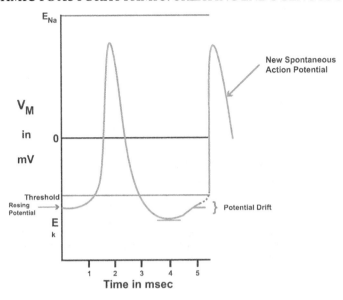

Figure 4.9: Potential drift after action potential leads to autorhythmic cell.

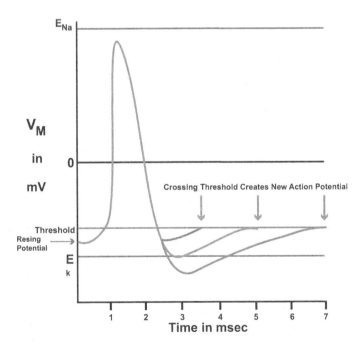

Figure 4.10: Changes in trough of action potential lead to frequency changes.

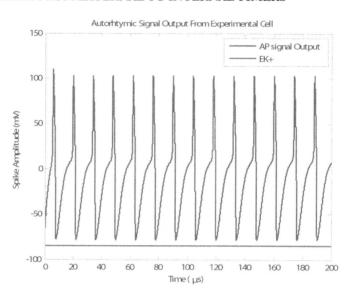

Figure 4.11: Autorhythmic cell generated by mathematical simulation of Hodgkin-Huxley model. (Modeled by G. Neusch and R. Gangulty.)

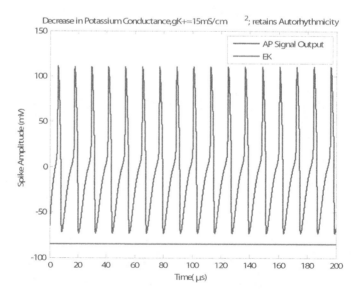

Figure 4.12: Decreasing potassium conductance increases frequency in autorhythmic neuron. (Modeled by G. Neusch and R. Gangulty.)

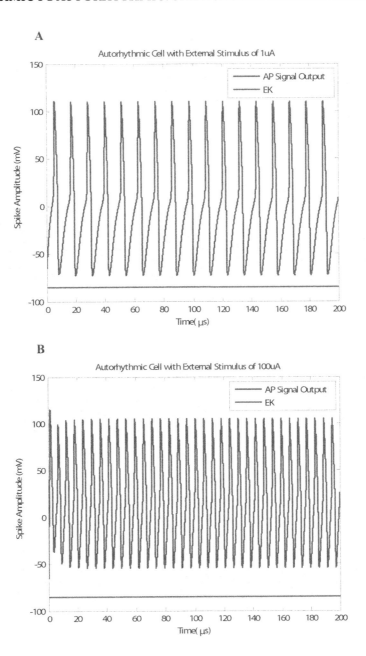

Figure 4.13: Autorhythmic neurons being subjected to an external stimulus. (A) the stimulus is 1 μA; (B) the stimulus was increased to 100 μA. (Modeled by G. Neusch and R. Gangulty.)

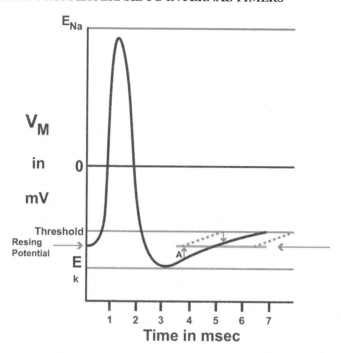

Figure 4.14: The effects of applying a stimulus forces the neuron to a fixed potential.

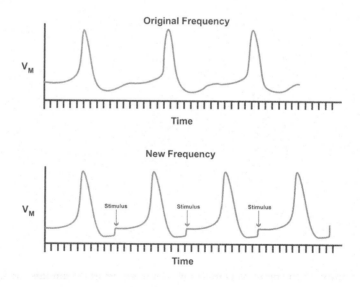

Figure 4.15: Effect of repeated stimulation on frequency in autorhythmic neuron.

If we continue along these lines, there appear to be two different methods of changing frequencies in autorhythmic neurons. First, a constant input could be applied, either stimulatory (depolarizing) or inhibitory (hyperpolarizing) which either raised the membrane potential or lowered it, respectively. Assuming that the sequence of changes in channel proteins which created the action potential in the first place could be maintained under such conditions, the result would be a consistent increase or decrease in spike frequency. These situations are displayed in Figures 4.13–4.15. In fact, this looks very similar to our previous discussion of applying a constant stimulus to a standard, non-autorhythmic neuron resulting in oscillations (Figure 4.8). Why bother with the evolution of autorhythmicity if the same result could be achieved without it?

The answer lies in the stability of the rhythm. There are significant differences in the two approaches to oscillations. In the case of a passive neuron (Figure 4.8), a constant stimulus must be applied to generate the output oscillation. This places the oscillation at the mercy of the input stimulus—no input, no oscillation. With an autorhythmic cell, the lack of an input merely results in the cell expressing its own inherent frequency. In the latter case, changes in input alter the *frequency* but not the *existence* of the oscillations. If oscillations are fundamental to the function of the system—as is the case with the heart—autorhythmic cells are a better evolutionary solution since they are buffered against changes in the environment that might otherwise result in a significant interruption or loss of rhythmicity. The effect of varying constant stimulus intensities is displayed in Figure 4.13.

Interestingly, there is a second method of changing frequency in autorhythmic neurons, in addition to a constant input. Consider the situation displayed in Figure 4.14. In this case, a single stimulus is applied at two different points in the neuron's autorhythmic cycle, points A and B. The effect of the stimulus in both cases is to reestablish the membrane potential at the theoretical resting potential (since this neuron is autorhythmic, it is never really at rest and thus does not have an actual resting potential). The interesting observation is that the exact same stimulus has very different effects depending on the time in the cycle at which it is applied. At point A, the stimulus moves the membrane potential *toward the threshold* and thus *decreases the time* it will take for the potential to cross threshold and reinitiate the next action potential. At point B, the same change in membrane potential moves the potential *away from threshold*, thus *increasing the time* it will take for the potential to cross threshold. Stimulating at point A *advances* the time for the next action potential while stimulating at point B *delays* the appearance of the next action potential. If one were to repeatedly stimulate the same neuron at point A, the result would be a neuron whose frequency of generating action potentials will have changed. The same is true for repeatedly stimulating at point B. The difference is that stimuli at point A will *increase* the frequency while stimuli at point B will *decrease* the frequency of action potentials coming from this autorhythmic neuron.

So, there appear to be at least two ways to alter the frequency of an autorhythmic neuron—a constant stimulatory or inhibitory stimulus or a repeated application of a single stimulus whose effects depend on the time the stimulus is applied. Either mechanism can change frequencies, so at this point, there appears to be no particular reason to choose one approach or the other.

4.6   ANOTHER BRIEF REVIEW

In summary, it appears that we have reached the same conclusion, albeit by a different route, as we did in Chapters 1–3. Living systems are inherently rhythmic, even at a cellular level. We have extended this conclusion, however, through the addition of endogenous rhythmicity—oscillations that are a fundamental characteristic of the living system and not imposed by any environmental cycle or even by an external stimulus of any kind at all.

In the heart, evolution has solved the question of generating robust rhythms with adaptability by developing a set of autorhythmic components—the nodes and conduction system—that impose a predictable oscillation on the entire organ. It is robust in that the rhythm can and does persist in the absence of any environmental stimulation. The system is adaptable in that frequency changes can be created to match physiological requirements. We investigated this phenomenon from the simple point of view of the generalized action potential characteristic of most excitable tissue, such as neurons and skeletal muscle cells. The cardiac myocyte (muscle cell) is a bit more complex and involves calcium currents as well as the aforementioned sodium and potassium currents. I will discuss this in more detail in a subsequent chapter. The important thing to note is the relative ease with which autorhythmicity could evolve from basic system parameters, either the feedback control system as described in Chapter 1 or the fundamental mechanism underlying action potentials, as described above. Additionally, there seem to be numerous means by which autorhythmic cells or systems could be adjusted to meet physiological and/or behavioral requirements. The latter observation brings us back to the issue with which we began the chapter—endogenous rhythms and the mechanisms by which these rhythms could aid in adjusting organisms to geophysical cycles.

4.7   SYNCHRONIZING ENDOGENOUS RHYTHMS TO ENVIRONMENTAL CYCLES

It appears that evolution does have the capability of creating endogenous oscillators that are both robust and flexible. If we use the heart approach of a few centralized pacemakers driving the entire system and apply it to rhythms associated with geophysical cycles, a new model emerges of a limited number of central pacemakers forcing a coherent rhythmicity onto the entire organism. Thus, internal synchrony is created through the dominance of a limited number of pacemakers, just as is the case with the heart. The question that remains is: How will external synchrony be achieved with the new model?

To achieve external synchrony actually requires that two conditions be satisfied. First, and most obvious, the frequencies of the biological rhythms and external cycles must match. In a model where environmental cycles were the sole driving mechanism for biological cycles, and under ideal conditions, such a match would be automatic. As we have seen, however, conditions are seldom ideal and a model depending solely on environmental drivers risks both external and internal desynchronization. The importance of temporal synchrony supports the evolution of pacemakers, capable of generating rhythms—and maintaining internal synchrony—despite minor environmental fluctuations. To ensure external synchrony, however, the frequency of pacemakers must be adjusted to match that of the dominant environmental cycle.

Frequency alone, however, is insufficient. As discussed in previous chapters, phase is just as important as frequency. Another example of this—at the scale of individual organisms—is presented in Figure 4.16.

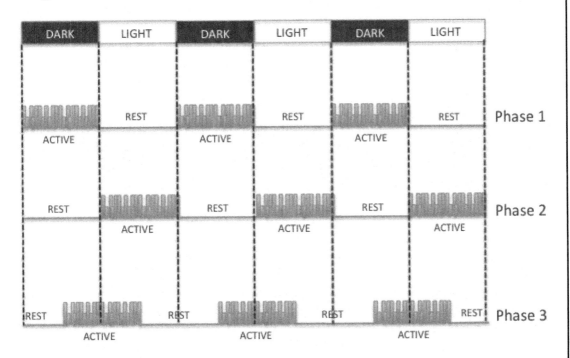

Figure 4.16: Three different rhythms with same frequency but different phase angles.

In Figure 4.16, three rhythms are displayed. The bars indicate the presence of activity. In looking at these three patterns, Phase 1 represents a nocturnal or night-active species, Phase 2 a diurnal or day-active species, and Phase 3 a possible crepuscular species which is active during the light-to-dark transition. In all three cases, the frequency matches that of the light/dark cycle but the phases are different, indicating a very different lifestyle. Thus, in order for adequate external synchrony, it is not enough that the frequency of the geophysical and biological rhythms match—the phase angle between them must be appropriate as well.

What does the term phase angle really mean? In Chapter 3, the phase angle was discussed in terms of hours. This limits any discussion of phasing to rhythms whose frequency revolves around that magnitude of time. However, any oscillation can be modeled as a series of repeating events which can be placed upon a circle consisting of 360 degrees. A twenty-four hour cycle, for example, could be modeled as a circle where each hour represents $15°$. Fifteen degrees on an annual cycle, on the other hand, represents about 24.34 *days*. Returning to daily rhythms, suppose we chose to use lights off and activity onset as corresponding phases between the light/dark cycle and biological rhythm respectively. In the case of Phase 1, the difference between the two would be 0°, i.e., they occur at the same relative point in both cycles. If we model the biological cycle as a clock of 360°, it appears that 180° are concerned with activity and the other 180° with rest. Using this information, the phase angle between lights off and activity onset in Phase 2 appears to be 180° and in Phase 3, it looks like a phase angle of about 90°.

The importance of the phase angle between the environmental and biological cycles is that it reflects the positioning of relevant biological events within the temporal environment. Recalling our previous discussion of the "on time" bird, the evolutionary advantage of biological rhythmicity was that it allows the bird to reliably predict when worms will be available and prepare both behaviorally and physiologically. The latter requires internal synchrony while the prediction itself necessitates external synchrony. The very act of making the prediction assumes a stable phase angle between some internal biological time and the worms' availability. In fact, there are two different phase angles in that situation. The phase angles between the bird's behavior and the day/night cycle and between the worms' availability and the day/night cycle must both be reasonably stable for the ecological timing to work.

It is therefore necessary to create two conditions for synchronizing an internal endogenous biological oscillation with an external geophysical cycle. First, the frequencies must match. Second, there must be a stable phase angle between the two rhythms. Can our simple model of an autorhythmic neuron handle these requirements or must we generate a new approach?

The autorhythmic neuron can be altered in two fundamental ways. First, the frequency can be made to change by applying a constant level of stimulatory or inhibitory input. While it is not clear whether or not a constant level of light or dark would change the frequency of a 24-hour biological pacemaker, it is clear that this cannot be how external synchrony is achieved. External synchrony demands a frequency match between two cycles, and a constant environmental input is not a cycle. What about the alternative approach—the time-dependent pulses?

Reexamining Figure 4.15 reveals a very interesting characteristic of the *time-dependent pulse effect*. The peak of the action potential is always in the same phase angle to the stimulus so long as the stimulus is given at the same point in the cycle. In other words, each time the stimulus is applied, an action potential results 5 time units later. Incredibly, not only does the stimulus alter the frequency, it generates a predictable and consistent phase angle from stimulus application time to action potential. If we consider this model a bit further, we can envision that a frequency match and phase angle stability could be generated by consistent application of a stimulus. In fact, the lower graph in Figure 4.12 shows exactly that—a stimulus given every 12 time units generates an action potential every 12 time units which would then always be 5 time units later than the stimulus! Thus, a pulsed stimulus which affects the phase of the neuron's cycle can be applied to create both a frequency match and phase angle stability between the cycle of stimulation and the rhythm in action potentials.

This is illustrated in greater detail in Figure 4.17. In that figure, the top graph represents the original (inherent) frequency of an autorhythmic (endogenous) neuron. The frequency can be determined to be approximately one spike every 13 time units, as indicated by the alternating red and blue lines at the top of the action potentials. Although there is certainly a phase (every rhythm has a phase), there is no phase angle, since there is no second cycle from which a phase angle could be calculated.

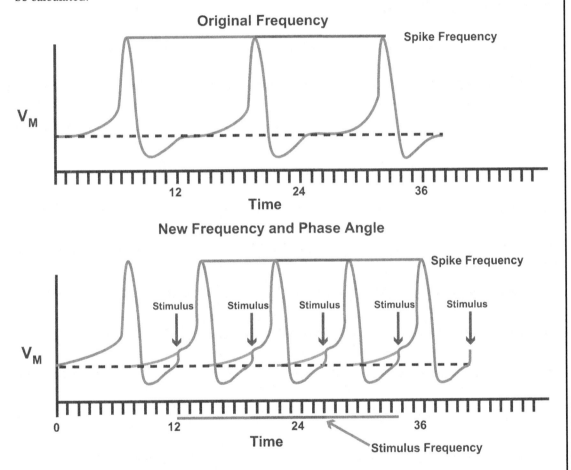

Figure 4.17: Frequency matching and phase angle stability in the autorhythmic neuron.

In the lower graph, a stimulus cycle with a fixed frequency has been applied to the cell. The stimulus rapidly accelerates the current change needed to reach an action potential, effectively cutting off a section of the neuron's rhythm (as indicated by the orange line). That part of the neuron's cycle, normally expressed without external stimulation, is eliminated when the stimulus is applied. As indicated with the alternating red and blue lines, the spike frequency has been changed from 1 spike every 13 time units to 1 spike every 8–9 time units. This matches the frequency with which the stimulus is applied to the neuron, i.e., the red and blue lines are of equal length for spike frequency and stimulus frequency. In addition, since there are two rhythms—one in action potentials and the other in external stimuli—a phase angle can be calculated between the two cycles. In this case, spikes always occur approximately 2–3 time units after application of the stimulus.

Of course, there are limitations to this model. A stimulus would not be effective if applied during the absolute refractory period and presumably would not have very much influence during the action potential itself. The best time for such a stimulus is during the rising phase of the membrane potential after the refractory period and before threshold is reached initiating the next action potential. If we reexamine Figure 4.14, we see that the frequency of an autorhythmic neuron could be either slowed (type B stimulus) or accelerated (type A stimulus) depending on the relative position of the stimulus along the potential rise. It also indicates some limitations on the amount of the frequency change that could be generated. Since there is only so much of a change that can be produced by each stimulus, this limits the size of potential phase shifts in the occurrence of the action potential. The size of the phase shift, in turn, determines the frequency change, since achieving a frequency match requires that the rhythm of stimuli match the action potential rhythm. That is, the phase shift would have to be such that each stimulus occurs at the same point in the potential rhythm.

As an interesting exercise, we may imagine creating a graph of the relationship between stimulus and phase change by applying a pulsed stimulus at different times in the neuron's cycle. Starting, for example, at time zero = threshold, an experimenter could apply a stimulus at time 0+1 time unit, then 0+2 time units, 0+3, and so on and measure the resulting phase change in the neuron's cycle. Assuming that pulses had little or no effect during the action potential and refractory periods, such a graph might resemble the graph presented in Figure 4.18.

In Figure 4.18, the top graph (A) represents the pattern of an autorhythmic neuron and the effect that the stimulus has upon the membrane potential. The blue line represents the potential the stimulus will generate when applied. In other words, the membrane potential will become that potential as a result of applying the stimulus. The red arrows indicate the effect of applying the stimulus. The area between the dotted lines is unaffected by the stimulus.

The bottom graph (B) plots the phase changes that result from applying the stimulus at various times in the neuron's autorhythmic cycle. Note that the graphs indicate that there are limits to which the cycle can be changed. Given the relationship between the magnitude of the phase change and the frequency of stimulus application, it appears that the frequency of this autorhythmic neuron can only be altered by a restricted amount. Since the neuron cycles once every 24 time units, it appears that it can be made to match a stimulus cycle as short as 1 cycle/12 time units or as long as 1 cycle/25–26 time units.

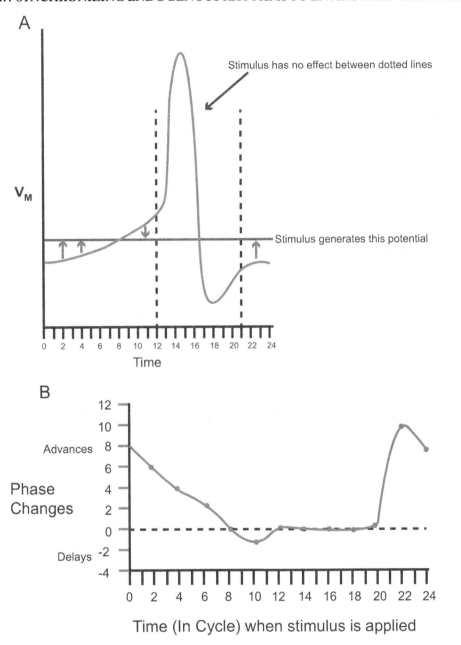

Figure 4.18: Relationship between time of stimulus application and phase change in autorhythmic neuron. A. Neuron and Stimulus Characteristics; B. Phase Change vs. Time of Stimulus Application Graph.

4.8   THE FINAL ANALYSIS

What have we learned and what can we conclude from our analysis to this point? We know that any adaptation must keep a reasonable level of internal and external synchrony. Internal synchrony maintains a temporal order necessary for the efficient functioning of a complex device and external synchrony places an organism in a temporal niche, allowing for adaptive prediction of external events associated with stable geophysical cycles. There seem to be three fundamental methods to generate the necessary rhythmicity:

1. Use of environmental cycles to drive biological rhythms;

2. Generation of oscillations through constant input (as with neurons);

3. Evolution of autorhythmic or endogenous "pacemaker" cells or tissues.

In Chapter 1, it was discovered that the feedback mechanisms in use to regulate many physiological processes are inherently oscillatory. As long as one component cycles, the entire system will oscillate in response. This led to the initial idea that environmental drivers associated with a geophysical cycle might generate biological rhythms. At first, it was supposed that each biological rhythm could be driven by a unique environmental cycle—temperature rhythms generating thyroid hormone rhythms, glucose oscillations generating cortisol rhythms, etc. It was quickly discovered that this approach would lead to internal desynchrony with variations in environmental cycles generating varying biological rhythms. Thus, this model proved unsatisfactory.

So, we then modified the model by reducing the number of environmental drivers to a single, stable geophysical cycle: the light and dark cycle associated with the Earth's rotation. This was an improvement over the previous model. However, problems remained in that the geophysical cycle of light and dark, while predictable in a strictly astronomical sense, is less stable when experienced from the point of view of an organism subjected to it. Organisms might be temporally isolated from light/dark cycles—such as in a burrow or cave—and thus loose contact with the environmental driver. Even when not strictly isolated from the cycle of light and dark, the light intensity experienced by organisms may differ under different conditions—cloud cover vs. a sunny day—and this raises the question of how organisms might interpret actual day length from a cloudy day to a sunny one (see Figure 4.3 and associated discussion). Also, there is a problem of adapting to other important cycles, such as seasonal changes. Thus, this improved model also seems unable to satisfy the twin conditions of internal and external synchrony.

At this point, we turned inward to examine an existing and familiar biological rhythm, the cardiac cycle. Oscillations within the heart must be internally synchronized and coherent while the heart rate must also be adaptable to the requirements of the body as a whole. The heart displays a strong oscillation with limited flexibility that might serve as an improved model for biological rhythms associated with geophysical cycles. To conduct this analysis, it was necessary to discuss the fundamental nature of membrane potentials, albeit in a very simple way, and how these potentials change in excitable tissue, such as neurons and muscle cells. In the process of this analysis, it was uncovered that neurons are naturally oscillatory. Given the correct level of consistent stimulation, a neuron will generate action potentials with a frequency characteristic of the specific neuron. This is yet another method of generating biological rhythms, one which does not require an oscillatory input.

At the same time, the nature of action potential generation could be easily modified to create a third mechanism for generating biological rhythms, which in this case required neither an oscillatory not constant input. In fact, no drivers are required at all. These cells are autorhythmic and generate rhythms as a results of mechanisms contained within their own structure. This is the third way of generating biological rhythms.

The latter mechanism turned out to be exceptionally interesting. Not only could such autorhythmic cells endogenously generate rhythms, those rhythms could be adjusted in both frequency and phase by the appropriate environmental stimuli. Of the two methods of generating adjustments, pulsed, periodic stimulation produced both a frequency match and stable phase angle between the rhythm of action potentials and the cycle of stimuli. A frequency match and stable phase angle between biological rhythm and the geophysical cycle are the conditions of external synchrony required for biological rhythms to be used to predict temporally stable environmental events. Thus, we have a new model for the generation of such biological rhythms—endogenous pacemakers.

Can this new model alleviate the problems associated with previous approaches without introducing new difficulties? Assuming, of course, that an autorhythmic pacemaker system can evolve with a period closer to 24 hours (or 28 days, 12 months, etc.) as opposed to the millisecond-seconds range of neurons and muscle cells, the key difference between pacemaker-driven rhythms and our previous models is the role played by the environment. In prior models, biological rhythms were driven by one or more environmental cycles. Thus, the very existence of the biological rhythms was dependent on the presence and detection of environmental cycles. The new model is analogous to the mechanism found in the heart—the main biological cycle is the result of an autorhythmic system. The role of the environmental cycle has shifted from that of *driver* to that of *synchronizer*. In the absence of any environmental cycle, the biological rhythm will persist although the frequency may change. Can this work in the real world of living systems and geophysical cycles?

The twin goals of biological rhythms are to maintain internal temporal order while fitting organisms into a time-varying environment in such a way as to allow preparation for predictable environmental changes. In other words, to maintain internal and external synchrony. From the point of view of those physiological and biochemical structures and systems downstream from the endogenous pacemaker(s), the difference between the two models of limited environmental drivers and limited pacemakers is undetectable under most natural conditions.

This situation is represented graphically in Figure 4.19. In the upper graph (A), the previous model is displayed. In this situation, biological rhythms are generated by the light/dark cycle and linked together through a common sensor/transducer. In the lower graph (B), the rhythms are actually produced by the pacemaker. The role of the light/dark cycle is to synchronize the internal biological rhythm system to the external geophysical cycle of day and night. From the point of view of the

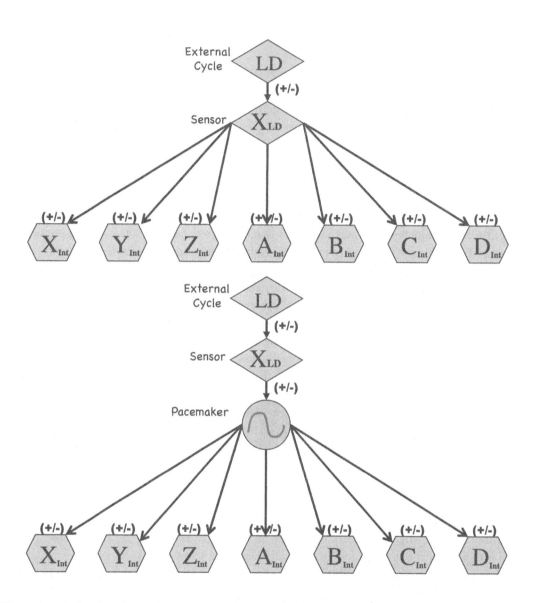

Figure 4.19: Graphical representations of models of biological rhythm generation. A. The light/dark cycle as the environmental driver generating biological rhythms. B. An autorhythmic model of biological rhythm generation where the light/dark cycle synchronizes the biological rhythms already in existence.

lowest level systems (A-D and X-Z), the source of timing information is external, whether from the separate pacemaker system or from the environmental cycle indirectly through the sensor/transducer (X_{LD}).

Thus, these internal systems are driven (at least in this model) by either the pacemaker or X_{LD}, and it is irrelevant from their perspective where the ultimate source of timing information arises. If this seems confusing, recall our discussion of systems and levels. Each low level system (A, B, etc.) can either be seen as a system in and of itself or as a component of the entire organism. If we take the former view, then every low level system is being driven into rhythmicity by forces external to itself. So long as those external forces act appropriately, there is no difference in the reaction of the low systems between the two models.

The key phrase here is "act appropriately." Recall that one of the problems with our previous model of using the LD cycle as the environmental driver of rhythms was that situations could be imagined in which the organism would not be exposed to the environmental cycle—in a cave, burrow, etc. The effect would be the same as if both the LD cycle and X_{LD} in Figure 4.19 were eliminated. In (A), there would no longer be timing information and the subsystems would become desynchronized from each other—internal desychronization. This is not an issue with the new model as shown in Figure 4.19B: all the subsystems would remain linked or coupled to the pacemaker whose rhythms would persist in the absence of external stimulation—just as is the case in the heart.

Wait a minute, you might protest! I am acting as if the rhythms would disappear instantly with removal of the LD cycle and that is clearly not the case. Rhythmic systems damp down over several cycles as Ken's mathematical models have demonstrated. Would that not be enough to maintain internal synchrony? My answer is: maybe. If you reexamine Figure 4.2, you will note that the rhythms lose amplitude very quickly and are eliminated entirely within a few cycles. The drop in amplitude means that any physiological signal indicating timing from X_{LD} is going to rapidly lose intensity. Think of this as analogous to a change in light intensity when the day/night cycle is being used to synchronize the main pacemaker. As the light intensity of the day decreases, there is going to come a point when the intensity is not sufficient to affect the pacemaker. The same is true for an internal, physiological signal linking the pacemaker with an internal rhythm. As the intensity of the signal drops, the ability of that signal to influence the subsystems will drop as well. It may last a cycle or it may not depending on the threshold of intensity required by the subsystems to indicate cycle timing. So, perhaps the better conclusion is that there is a risk of internal desynchronization with environmentally driven rhythms which is virtually eliminated when using autorhythmic pacemakers.

What about the light intensity effects described in Figure 4.3? Again, an autorhythmic system has an advantage over environmentally driven rhythms. A pacemaker will run at a frequency even if not stimulated. If the day is overcast and the system is not stimulated, an autorhythmic system will continue to cycle and the organisms will organize its behavior and physiology the next day at approximately the same time as the previous one. This allows evolution some latitude in dealing with the signal sensitivity problem previously discussed. Recall that the basic problem is that if an environmentally driven system were too sensitive, it would vary in frequency continuously, creating difficulties in maintaining both internal and external synchrony. If the system is not sensitive enough, the environmental cycle would remain undetected and both internal and external synchrony would be lost. With an autorhythmic system, the latter problem is considerably reduced. Even if the environmental cycle were undetected for a few cycles, a pacemaker system maintains internal synchrony and if the pacemaker frequency were reasonably stable, external synchrony could be maintained as well. A considerable improvement in adaptability is provided by endogenous rhythmicity.

So, it appears that autorhythmic systems confer significant adaptive advantages to organisms in terms of synchronizing to daily rhythms. What about the seasonality problem? The difficulty described in Chapter 3 in regard to organisms adjusting to seasons involved the subtle changes in the timing of dawn and dusk and the photoperiod changes associated with the Earth's revolution about the Sun. The problem was that organisms able to move across latitudes will experience artificial seasons, since depending on where they are, they will have different dawn/dusk times and, as a result, photoperiod lengths. *Perception is reality for most organisms whether actually correct or not.* This could be considered another sensitivity problem, in this instance associated primarily with photoperiod length. A pacemaker system with an annual cycle (1 cycle/365 days) would buffer the organisms against this problem in an analogous fashion to a daily pacemaker system buffering an organism against changes in the cycle of daily light intensity. Autorhythmicity seems to have a rather considerable evolutionary advantage over rhythm systems solely driven by environmental cycles.

4.9   CHAPTER REVIEW

In this chapter, we further analyzed a model of environmentally driven biological rhythms using the light/dark cycle as the sole driving oscillation. Despite the improvements afforded by this approach over multiple environmental cycles driving multiple biological rhythms, problems associated with maintaining both internal and external synchrony remained. Using the heart as a model system, we investigated the basic mechanism of action potentials only to discover a new form of biological rhythm, one that could be initiated with constant, rather than oscillatory, environmental input. Further investigation led to the conclusion that relatively minor changes in the basic parameters of action potentials could lead to the evolution of autorhythmic neurons. These neurons could be adjusted by environmental cycles to create both a frequency match to the environmental cycle and a stable phase angle between the two oscillations. Insofar as these are the conditions necessary to create external synchrony, we turned our attention to a new model of biological rhythms associated with geophysical cycles, autorhythmic or endogenous pacemakers.

Using a simple model of such pacemakers, several of the disadvantages associated with environmentally driven rhythms are reduced or eliminated entirely. Pacemaker systems are more rhythmic, stable, and show greater ability to maintain internal and external synchrony in the face of environmental perturbations. From a modeling and evolutionary standpoint, pacemaker systems would seem to be the preferred approach to maintaining temporal coherence and adaptability.

The model of an autorhythmic neuron can be used to generate both a frequency match and stable phase angle with a pulsed, stably cycling stimulus. A curve of phase changes associated with the timing of the pulse can be created—a kind of *phase response curve*—which shows how such an autorhythmic neuron would react to various frequencies of pulsed stimuli. An interesting feature of the model is that there is a limited range of frequencies of pulsed stimuli to which such a neuron could synchronize. These two features—a phase response curve and *limited range of synchronizing frequencies*—were generated by the model itself and it remains to be seen if the features are generally applicable to other types of biological pacemakers.

REFERENCES

[1] Mathews, G. (1998). *Neurobiology: Molecules, Cells and Systems*. Blackwell Science: Malden, MA. 95

[2] Michael, J.A. and Rovick, A.A. (1998). *Problem Solving in Physiology*. Benjamin Cummings: New York. 96, 104

CHAPTER 5

Clock Tech III Rise of the CircaRhythms

Overview

In this chapter, we explore the implications of the evolution of *circarhythms—endogenous biological clocks* capable of being synchronized to geophysical cycles and generating both *internal* and *external synchrony*. We will review certain general characteristics of such clocks using *circadian* (near 24 hour) rhythms as the model. We will then return to the *empirical proof-of-concept* concerning the importance of biological rhythms begun in Chapter 2. In Chapter 2, we were unable to determine the effects of rhythm disruption on human health and performance insofar as we were not certain as to the physiological nature of the rhythm generating process. Are observed daily rhythms merely driven by daily environmental cycles? Or are such rhythms based upon endogenous biological clocks which are synchronized (entrained) by environmental cycles? In this chapter, we proceed on the basis that *most daily rhythms are true circadian cycles*. Under these circumstances, *rhythm disruption* can be achieved through various means, including *rapidly changing environmental cycles*. We investigate the effects of circadian *rhythm disruption in both humans and animals*, using both *correlational* and *true experimental approaches*. We will concluded that circadian *rhythmicity is fundamental* to human health, well-being, and performance, supporting the hypothesis that *biological rhythms in general are critical to both physiological and behavioral functioning*. Finally, we compare and contrast the role of circarhythms, which function as *biological clocks*, with another kind of timer—an *interval timer*. Such timers play a different, albeit complimentary, role in creating *temporal organization* in biological systems.

What is time? The shadow on the dial, the striking of the clock, the running of the sand—day and night, summer and winter, months, years, centuries—these are but arbitrary and outward signs, the measure of time, not time itself. Time is the life of the soul.

-Henry Wadsworth Longfellow

He who gains time gains everything.

-Benjamin Disraeli, 1ˢᵗ Earl of Beaconsfield

5.1   STARTING FORWARD BY LOOKING BACK

Let us review what we have learned so far in this book. First, it appears that *biological systems are inherently oscillatory*. In Chapter 1, we found that the more or less standard model of endocrine signaling using negative feedback and multiple setpoints began to cycle as soon as any component of the model began to oscillate. It did not matter whether the component was an external or internal stimulus (represented by X in the model), hormone secretion, liver metabolism, or any other factor—if one part of the system were made to cycle, the entire system became rhythmic in response. This observation was extended in Chapter 4. The standard model of the neuron action potentials results in a neuron that oscillates in response to a constant input stimulus. This point is worth reiterating: for the average neuron, there need not be an oscillating input to generate an oscillating output. The very nature of the activation process itself generates cycles.

We then turned our attention to another aspect of oscillations, the external environment. There, we again found a situation dominated by cyclic phenomena. Geophysical cycles, such as the Earth's rotation on its axis and revolution around the Sun, manifest themselves as sequences of alternating light and dark and seasonal changes. These changes create a *predictable, time–varying ecology into which organisms must adapt*. As a consequence, there are temporal ecological niches into which every organism must fit if it is to thrive.

Furthermore, we were able to discover that oscillations create a type of *systemic cascade*. By this I mean that in any sufficiently interconnected system, once one component begins to oscillate, the entire system becomes oscillatory. This is an extension of what we found with our simple endocrine model. Make one component cycle and the entire system cycles as a result. Organisms and their external environment can be viewed, on evolutionary grounds, as a vast interconnected system. Cycles in ecological systems are a natural consequence of the existence of geophysical cycles. Once ecological systems oscillate, organisms will display behavioral oscillations. Behavioral oscillations will lead to physiological oscillations, and physiological oscillations will lead to biochemical oscillations. Living systems are a vast collection of oscillations, rhythms, and cycles.

There is another way we can reach the same conclusion. Chapter 1 discusses the need for internal timing mechanisms in any complex, goal-driven device. Computers, automobiles—it does not matter. All such *devices require an internal temporal order in order to function properly*. Most such timers are oscillatory in nature since the set of activities being organized tends to repeat itself. To be effective, a spark plug cannot fire just once—it is necessary to coordinate a repeating cycle of multiple plug firings at the correct rate to generate the motive force to move a car. Living systems are also complex and goal-driven—evolution provides the goal as the preservation and propagation of genetic material. Thus, as complex, goal-driven devices, all living systems must have internal timing systems. Moreover, since living systems engage in a considerable amount of repetitive behavior and physiology (metabolic cycles, for example), it makes sense that such internal timers should be oscillatory in nature.

The "on-time" bird of Chapter 1 illustrates the critical importance of accurate timing along with the concepts of internal timing and the systemic cascade. Assume that the day/night cycle creates a circumstance where worms are available only at certain times of day. As a result, the best strategy for birds feeding on those worms is to arrive at, or just before, the time worms become available. Thus, the geophysical cycle begets a cycle in worm behavior which then, in turn, begets a cycle in bird behavior. What about the bird's internal physiology? For the bird to absorb food, digestive enzymes must be synthesized, blood flow changed, absorptive processes initiated, etc. All these processes require energy. You may have noticed that you feel warm after a big meal. This is called *postprandial hyperthermia* and is the result of energy being directed to the processes of digestion and selective absorption. Some of the processes, such the synthesis of digestive proteins, takes both time and energy, so it may be a more successful evolutionary adaptation to only produce these proteins around the time when they will be needed rather than continuously (a "just-in-time" approach). Thus, the digestive system of the on-time bird will react to the periodic appearance of worms as a food item by periodic alterations in protein synthesis. There are other preparations within the digestive system as well that can prepare the system for food to increase the efficiency of digestion and absorption. Detoxification enzymes in the liver should also be available for the periodic appearance of food that may contain such toxins. And so, the cascade of evolutionary adaptation extends from the day/night cycle to worm behavior to bird behavior to bird physiology and on down to bird biochemistry.

This example also illustrates the *necessity of maintaining external and internal synchrony*. If any part of the cascade is timed improperly, the system loses efficiency. If birds arrive on the feeding grounds when worms are not available, they can starve, even if their internal systems are all properly synchronized. On the other hand, if a bird ingests significant levels of toxins and the enzymes needed to deactivate those toxins are not available, the bird could suffer internal damage despite having arrived on the feeding grounds at the right time. Thus, both internal and external synchrony must be maintained for the optimum outcomes to be achieved.

In Chapter 3, we attempted to determine how timing systems might have evolved to solve the twin requirements of internal and external synchrony. The initial approach was to assume that every physiological system is driven by the environmental cycle to which the system corresponds. An external cycle of heat and cold driving the rhythm in thyroid hormone release is an example of the approach. After some analysis, we concluded that such a system would be too variable and likely to generate internal desynchronization. A better system was needed.

The next approach we considered was to limit the number of driving external cycles to those with the greatest stability and, thus, predictability. The new model used only the light/dark cycle generated by the Earth's rotation as the sole external driver for all internal rhythms. While this approach mitigated the problem of maintaining internal synchronization, a new problem arose involving signal strength and sensitivity. The natural day/night cycle can vary in the timing and intensity of light available to an organism. Sunrise on a cloudy day reaches a specific light intensity at a different time than on a sunny day, leading to problems in maintaining external synchrony. If an organism were too sensitive to light intensity, day-to-day variations in the timing of light intensity will cause rhythms to fluctuate erratically. If the organism were not sensitive enough, timing signals might be missed altogether. So even though the new model was an improvement over the previous one, it did not seem adequate to the task of maintaining consistent internal and external synchrony.

This brings us to the last chapter wherein we examined the heart as a potential model system. In the heart, a limited number of cells become *autorhythmic*, generating a baseline cycle. The cells are collected into centralized structures, the AV and SA nodes, which enforce a consistent internal rhythm upon the rest of the organ. These cell associations—called *pacemakers*—not only generate a rhythm without the necessity of a stimulating rhythm external to the structure but also provide central control points through which adjustments to the rhythm can be accomplished. The latter circumstance avoids the problem of having to adjust every cell individually, which, as we have seen before, leads to a high potential for internal desynchronization.

We then attempted to uncover a mechanism by which excitable tissue might become autorhythmic. Rather than analyze the more complex cardiac pacemaker cells, we started with a simple model of the neuron (nerve cell) to examine. Using a standard model of action potentials, we found that minor changes in the parameters of that model led to an autorhythmic output. Moreover, these model nerve cells could be adjusted in terms of frequency through either constant input or by timed pulses. In the latter case, both phase and frequency can be adjusted to match an external cycle. A frequency match and consistent phase angle between a biological rhythm and an environmental cycle are the exact requirements needed for external synchrony. It would seem that a limited number of controlling rhythms driving the entire physiological system (as found in the example of the heart) can maintain internal synchrony while simultaneously providing a means for external synchronization. This means that *a model of a limited number of autorhythmic pacemakers linked to a stable external cycle would appear to be an appropriate mechanism for maintaining external and internal synchrony.*

That brings us to this chapter where we will examine this new model in greater detail.

5.2   CIRCARHYTHMS

What is a circarhythm? *Circa* is Latin for "about" or "approximately." Therefore, a circarhythm must approximate another rhythm in some parameter. The parameter in this case is frequency (or period, period being just the inverse of the frequency). Thus, a circarhythm approximates the frequency of another cycle.

So far, so good. The problem is that this definition does not actually mean anything. After all, all rhythms have a frequency, and thus all rhythms approximate a frequency. In that case, maybe we should refer to *any* biological rhythm as a circa-something cycle. For instance, could we refer to neurons as displaying circa-millisecond rhythms? Is the cardiac cycle a circa-second rhythm? If so, what is the purpose of adding the prefix "circa" to anything? Indeed, maybe we would be better off just providing the frequency of the rhythm in question and not bothering with the term circa at all.

Well, *circa* does mean approximately and thus adds an element of variability to the frequency that might otherwise be missed. Using the word stresses the importance of the natural variation in the frequency. A cardiac cycle is not always 0.8 seconds and so the term *circa* may be justified. However, in this text, we are going to use a far more restrictive definition. For our purposes, a circarhythm has the following characteristics:

1. A circarhythm is a *biological rhythm*;

2. The biological rhythm is *endogenously generated*, that is, generated autorhythmically by the cell(s), tissue(s), or organism(s) in question;

3. The biological rhythm *approximates the frequency of a geophysical cycle*;

4. The biological rhythm *can be synchronized to that geophysical cycle* resulting in both a frequency match and stable phase angle between the biological rhythm and geophysical cycle.

Not all researchers in the field of biological rhythms will agree with this definition, considering it too restrictive. In the interests of clarity, however, we will at least begin with this definition. If it requires expansion later, we will reevaluate it.

Figure 5.1 is a version of the previous figure from Chapter 2. The globe represents the geophysical cycle to which the biological rhythm is to synchronize. Any such cycle capable of synchronizing a circarhythm is called a Zeitgeber (German for "time-giver") and is designated $\mathbf{Z}$ in the figure. The blue circle represents the autorhythmic central pacemaker ($\mathbf{P}$) while the CircaRhythms ($\mathbf{R}$) are the observable output rhythms in physiology and behavior. The arrow labeled $\mathbf{C_{zp}}$ represents the combination of a sensor for the geophysical cycle (a light sensor for the day/night cycle, for example) and the physiological link between the sensor and the pacemaker. In mammals, for example, the retina contains light sensors and retinal ganglion (nerve) cells that transmit lighting information via the retino-hypothalamic tract to the appropriate structure. This structure is a region of the hypothalamus called the suprachiasmatic nuclei (SCN), which contain the pacemakers. A second link, $\mathbf{C_{pr}}$, transmits timing information from the pacemaker to whatever structures generate the physiological and/or behavioral activity. Again using mammals as an example, information from the SCN is transmitted to another hypothalamic nucleus down to sympathetic neurons in the spinal cord, from there to post-ganglionic neurons in the superior cervical ganglion and then back up into the brain to connect to the pineal, generating a robust rhythm in melatonin secretion (Refinetti[73]).

What are the operating characteristics of a pacemaker capable of controlling a circarhythm? Using the autorhythmic neuron as a kind of prototype, there are several characteristics that might be expected of a circa-pacemaker.

First and foremost, the natural frequency of a circa-pacemaker must be fairly close to that of the geophysical cycle to which the corresponding biological rhythm would synchronize. Synchronization (called *entrainment* under these circumstances) requires both a frequency match and phase angle stability. In the case of the autorhythmic neuron, we learned that this was best achieved by timed pulses from an external source. The result was a graph of phase shifts associated with the timing of the pulses within the action potential cycle as shown in Figure 5.2 (reproducing Chapter 4, Figure 4.15b). One characteristic of that graph, or *phase response curve*, was that the magnitude of possible phase shifts was limited—the action potential rhythms could only be shifted by certain amounts. If you

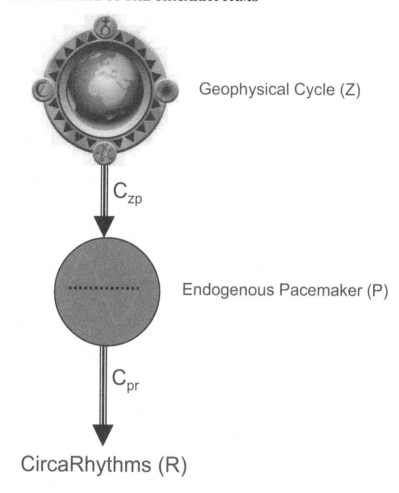

Geophysical Cycle (Z)

C_{zp}

Endogenous Pacemaker (P)

C_{pr}

CircaRhythms (R)

Figure 5.1: Minimal model for a circarhythm. (Globe courtesy of http://www.webweaver.nu/clipart/earth.shtml.)

examine the effects of timed pulses in Figure 5.3 (reproducing Chapter 4, Figure 4.14), you will see that the magnitude of the phase shift is directly related to the frequency achieved by the neuron. The greater the magnitude of the phase shift, the greater the difference between the natural frequency of the autorhythmic neuron and the new frequency generated by the cycling stimuli or pulses. Thus, if the magnitude of the phase shifts is limited, then the range of frequencies that the autorhythmic neuron can achieve will also be limited.

Since the whole point of the model is to create a situation in which entrainment with a geophysical cycle could be achieved, it makes sense to predict that the circa-pacemaker's natural

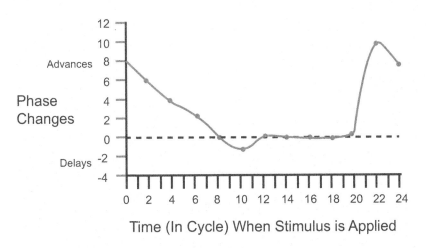

Figure 5.2: Phase change vs. time of stimulus application graph.

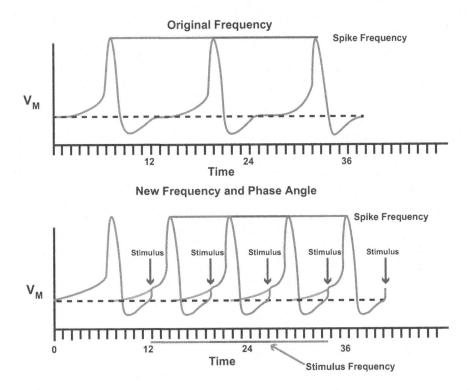

Figure 5.3: Frequency matching and phase angle stability in the autorhythmic neuron.

frequency must be close enough to that of the geophysical cycle to make entrainment possible. Another way of saying this is that the geophysical cycle should lie within the circa-pacemaker's *range of entrainment*. Later, we will discover that there is another, even more fundamental reason for a range of entrainment, but this will do for now.

What about the effect of a constant input stimulus? In the simple autorhythmic neuron model, changes in the level of a constant input stimulus led to frequency changes. This opens up the possibility that our model pacemakers might show the same or similar characteristics. It may be a little difficult to imagine an adaptive application for an effect only likely to arise under laboratory conditions. However, modern cities do generate significant levels of light, 24 hours a day, 7 days a week. It would be interesting to see what effects this constant light exposure might have on living circa-pacemakers.

Finally, is there anything else that could be added to our discussion of circa-pacemaker organization from an evolutionary perspective? In Chapter 2, I justified extrapolating research results from animals to humans based upon the ubiquity of daily rhythms. The concept underlying this justification comes from the modification by descent aspect of natural selection. *The more widespread a given adaptation is throughout existing organisms, the further back in time it is likely to have originated and the more likely it is that existing organisms share a common ancestor for that adaptation.* Daily rhythms appear in virtually all eukaryotic organisms studied and in some prokaryote species as well. Therefore, it can be argued that the adaptation is truly ancient and that many existing species inherited this fundamental mechanism from a common ancestor. In fact, there appear at least two separate origins for this adaptation—one for eukaryotes and another for prokaryotes—but the argument for the early development of daily rhythmicity is still valid.

What if, however, the adaptation represents more than merely the ability to express daily oscillations, which could be generated by environmental cycles? What if the adaptation is actually the capacity to generate pacemaker-driven circadian rhythms? In that case, circadian pacemakers become a fundamental characteristic of all existing eukaryotic organisms *and their ancestors*. This has rather profound implications for the model presented in Figure 5.1.

The argument proceeds as follows. All eukaryotic organisms have circadian rhythms and therefore either contain circadian pacemakers or act as pacemakers themselves. This even includes existing single-celled eukaryotes, such as *Euglena*. This pacemaker capacity was inherited from ancestral organisms. Just how far back does it extend? If all eukaryotic organisms truly share this adaptation, both single-celled and multicellular, then the adaptation most likely originated prior to the evolution of multicellular organisms. This implies that all cells may currently have the potential—inherited from their evolutionary ancestors—to express circadian rhythms.

How does this evolutionary legacy affect the organization of circadian rhythms within multicellular organisms? If we return to the heart as a model of autorhythmicity, it is not the case that the nodes or pacemakers alone have the capacity for generating rhythms. Other components of the heart are also rhythmic. Under normal conditions, heart rate reflects the frequency of the sinoatrial (SA) node because that pacemaker dominates (entrains) the rest of the autorhythmic components by having a higher frequency. In essence, the SA node behaves like the pulsed stimulation displayed in Figure 5.3, and it entrains the rest of the heart to its own frequency. Thus, the heart consists of multiple rhythmic components entrained to a single frequency by a master pacemaker.

We may envision a similar organization for circadian systems within multicellular organisms. A few master pacemakers act as central control points and interfaces between internal and external rhythms. These pacemakers act to maintain rhythmic coherence within the organisms by being entrained to selected environmental cycles (Zeitgebers) and then entraining internal oscillators to their own dominant cycles. The difference between this model (Figure 5.4) and that of Figure 5.1 is actually the same as between the model of rhythms presented in Figure 3.4 (Chapter 3) and that proposed in Figure 5.1. In the model described by Figure 3.4, environmental cycles acted as drivers, as *creators* of biological rhythms whereas in Figure 5.1, the role of environmental cycles is to act as *synchronizers*, as Zeitgebers for pre-existing biological rhythms. This new model for circadian organization envisions the role of the master pacemaker(s) as *internal entraining agents* (*internal Zeitgebers*), which act more like the conductor of a rhythmic symphony and less as the player of every instrument. This is illustrated in Figure 5.4.

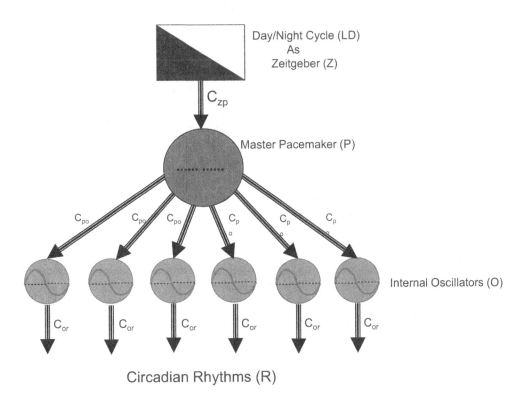

Figure 5.4: Suggested minimal model for circadian organization in multi-cellular organisms.

In Figure 5.4, oscillations occur at multiple levels and must be coordinated with each other for the system to function properly. The dominant geophysical cycle is that of the Earth's rotation, manifesting itself as an alternating cycle of light and dark (LD cycle). Within the multicellular organism, there is a sensor system for detecting the LD cycle and conveying that information (C_{zp}) to a master pacemaker (P) in order to entrain the master pacemaker to the external environment. However, the situation within the organism is more complex than previously envisioned and there are a number of internal oscillators (O). To achieve the required internal and external synchrony, the master pacemaker must synchronize (entrain) the internal oscillators to itself and through this entrainment, synchronize them to the external LD cycle as well.

The preceding provides one model for coordinating rhythms that have a frequency of approximately 1 cycle every 24 hours. The situation gets even more complex as soon as we consider that there are multiple rhythms operating at various frequencies which also must be coordinated together. We will explore some of these issues and rhythms in subsequent chapters as well as in Volume 2 of this series.

This is all well and good, but can we claim that all eukaryotes have circadian, rather than merely daily, rhythms? From the models we have examined in previous chapters, it is pretty clear that one has to keep an organism isolated from geophysical rhythm—such as a light/dark cycle—for quite some time in order to determine if the rhythms are circadian or merely damping down of rhythm generated by environmental cycles. In fact, Enright[22] recommended that a *phase reference point* (a characteristic of a rhythm to which it reliably returns, such as activity onset or peak level) be chosen and a rhythm be kept in constant conditions until the phase reference point moved through the entire 24 hours of clock time. For example, if the phase reference point were activity onset and the rhythm frequency were 1 full cycle every 25 hours, it would take 24–25 days of temporal isolation before you could be certain that the rhythm was truly circadian in nature. This is illustrated in Figure 5.5.

Clearly, it is far easier to establish the universality of *daily* rhythms than true *circadian* rhythms. Observations of species under natural conditions can establish the former, while investigating the latter requires strictly controlled conditions within a laboratory setting. How can we be absolutely certain that daily rhythms observed in free-living species are actually expressions of an underlying circadian system without laboratory verification?

The short answer is that we simply cannot be absolutely certain. However, there is a way to provide support for the idea, although it may seem at first like circular thinking (no pun intended!). In Chapter 2, I used the ubiquity of daily rhythms as providing an evolutionary basis to justify extrapolating animal research data to humans. We will now turn this argument around in the following way. Suppose a large number of very different species are examined and it is determined through rigorous application of the proper experimental procedures that the daily rhythms these species exhibit are, in fact, circadian. We can then entertain the possibility that most, if not all, observed daily rhythms have an underlying circadian structure. This argument may be reinforced through molecular biology if we can demonstrate that the molecular mechanisms underlying the structure

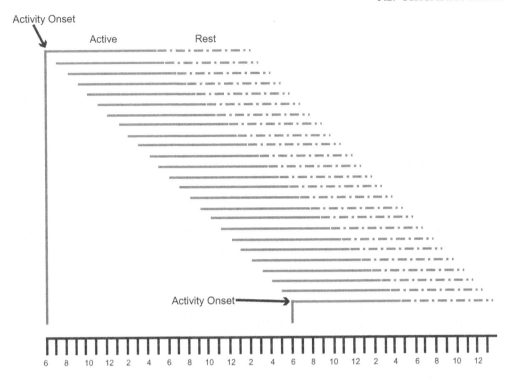

Figure 5.5: Establishing a circadian rest/activity rhythm by observing activity onset occurring at all clock times during exposure to constant environmental conditions.

of circadian pacemakers in widely divergent species show characteristics that indicate a common ancestor in the distant evolutionary past. In fact, the latter does indeed appear to be the case,[62] and circadian rhythms have been extensively documented not only in intact species but also in tissue and organ systems[75]. Thus, we can reasonably justify proceeding with the idea that not only daily, but more specifically circadian, rhythms are a ubiquitous characteristic of eukaryotic organisms.

What does this hypothesis do for us in our investigations into biological rhythms and the role played by such rhythms in human health and well-being? It allows us to return to a question in Chapter 2 that we were previously unable to answer: What is the underlying nature of daily rhythms? Recall that we needed an answer to this question to provide evidence for Proposition 2.3: *Disruption of a biological rhythm will generate deleterious effects on the biological system*. If the underlying nature of daily rhythms is a circadian organization of pacemakers and oscillators as displayed in Figure 5.4, we are now in a position to determine the truth (or falsehood) of Proposition 2.3.

Before proceeding, I want to clarify something about the nature of circarhythms. The main geophysical cycle to which most organisms must adapt is the day/night cycle, leading to the evolution of circadian rhythms. There are, however, other geophysical cycles of great adaptive significance to various organisms, including tidal rhythms (circatidal), lunar cycles (circalunar), and seasonal variations (cirannual). However, whether an organism expresses these latter cycles is dependent upon the organism's particular ecological niche whereas circadian rhythms appear to be universal among eukaryotic organisms. Because of their universality, for the time being, we will focus on circadian rhythms to test Proposition 2.3.

Proposition 2.3 *Disruption of a biological rhythm will generate deleterious effects on the biological system.*

In addition to the underlying circadian structure, we also need to consider the nature of the Zeitgeber. The most common environmental cycle known to synchronize circadian rhythms is a cycle of alternating light and dark[68]. There are other environmental rhythms known to act as Zeitgebers, but light/dark (LD) cycles appear to be the most common[7].

Armed with both our models and our observations, we can now turn to the question of disrupting circadian rhythms. Without any further information about the exact biochemical or physiological nature of the system, we can envision three possible means for disruption:

1. The application of a light pulse of sufficient strength at single point in the circadian cycle;

2. Exposure to a constant level of high intensity light; and

3. Exposing the organism to constantly shifting LD cycles.

How did we get from circadian oscillators and LD cycles to those three mechanisms for disrupting circadian rhythms? Hopefully, I can clarify the connections. Let us begin with #1—the single light pulse. This is really an application of a general phenomenon involving any kind of oscillating system. If the right kind of stimulus is applied at just the right point in time with just the right force, any oscillation can be stopped. As a physical example, consider a pendulum, which displays periodic motion by swinging back and forth (Figure 5.6). Without going into too much detail, the pendulum is swinging with a certain force as it reaches the bottom of the swing. This force has both a magnitude and a direction, as indicated by the arrow. If you were to hit the pendulum just as it reached the nadir of the swing with a force equal in magnitude but in the exact opposite direction, the two forces would sum and the pendulum would stop cold. An often tragic biological example is the application of physical force to the chest, and the heart contained therein, which can result in cessation of the cardiac cycle.

Although we cannot be sure that light is the correct stimulus for the circadian oscillator, at this point in our discussion, it is the only stimulus of which we are currently aware. Thus, we can hypothesize that a pulse of light, given at the right time and in the right amount, might cause the oscillator to stop.

What about constant, high intensity light? How does that work? There are really two questions here—why constant light and why high intensity light? To understand this idea, consider the drastically simplified version of Chapter 1's original model for an oscillating system as presented in Figure 5.7. In Chapter 1, we were able to demonstrate that a model based upon negative feedback

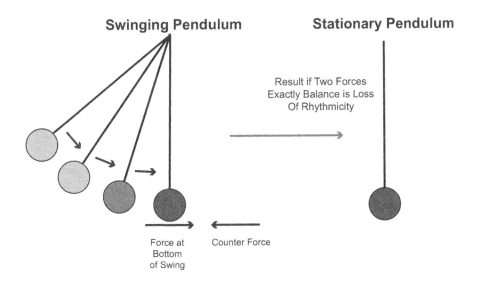

Figure 5.6: Stopping an oscillating system.

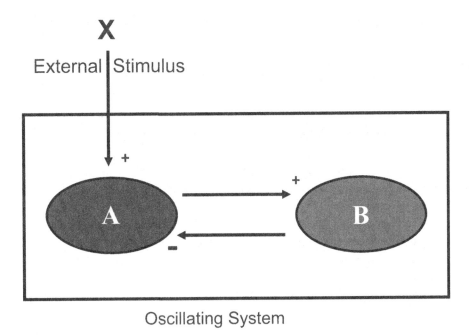

Figure 5.7: Simple oscillating system with an external stimulus.

was inherently oscillatory. As long as one of the major parameters was cycling, the entire system had a tendency to cycle. An interesting result of the mathematical model used to simulate the model was that a single pulse of the external stimulus, X, was sufficient to initiate oscillation. Given the inherently oscillatory nature of feedback control systems, and the widespread use of such systems in biological organisms, it is not too great a stretch of the imagination to hypothesize that the circadian oscillator would be based on some sort of feedback control system. Natural selection is nothing if not conservative, and there is no reason to suspect the evolution of an entirely new approach to time measurement if a satisfactory mechanism was already available. Figure 5.7 is a very simple representation of such an oscillating system.

Let us examine the impact of X on the system. From the results reported in Chapter 1, we already know that if X is rhythmic, the entire system will become rhythmic. Further, a single pulse can initiate cycling in the system even without further stimulation from X. In other words, once cycling, the system no longer needs external stimulation to maintain rhythmicity. What happens if the input from X is constant, however?

As with most of biology, the answer to that question is: it depends. Specifically, the result depends on the intensity of the positive (+) and negative (−) stimulation and the actual level of X. At low levels of X, X stimulates A which stimulates B which then inhibits A. Thus, the system begins to oscillate so long as the negative feedback exerted by B exceeds the stimulation provided by X. As X increases, the stimulation of A may increase to the point where it exceeds the maximum negative feedback provided by B. At that point, cycling will break down.

We can test this approach with our two existing models for biological rhythms. First, we can examine the effects of a constant level of stimulus in our endocrine feedback model. As a reminder, the model itself is provided in Figure 5.8.

Figure 5.9 shows a constant high intensity stimulus applied to the endocrine gland in an oscillating system. In this case, the system is made autorhythmic by means of a cycling intra-endocrine signal. Due to the intensity of the stimulus, the enzyme works within a smaller dynamic range and eventually becomes saturated (see the lower right-hand graph). Hormone production is increased until it too becomes saturated. What was an oscillating system at a lower level of stimulation (or no stimulation at all) ceases to have a cycling output in most of its components. Note, however, that the intra-endocrine signal continues to oscillate even when the rest of the system is non-cycling. Thus, the system may be damped but yet the original oscillating component remains rhythmic.

We can also apply the same idea to our model of the autorhythmic neuron. Figure 5.10a displays the basic action potential pattern of an autorhythmic neuron and Figure 5.10b shows the result of applying a constant stimulus of 340 μA which causes the rhythms to damp out.

Thus, our two questions are answered. First, there must be stimulation by *constant* light since any periodic input will simply generate or maintain rhythmicity in the system. Second, there needs to be a relatively *high light intensity* input to overcome the feedback components of the oscillator. The actual threshold for high intensity will depend on the specific system components in the organism being studied. We are only considering light as the external stimulus at this stage since it is the

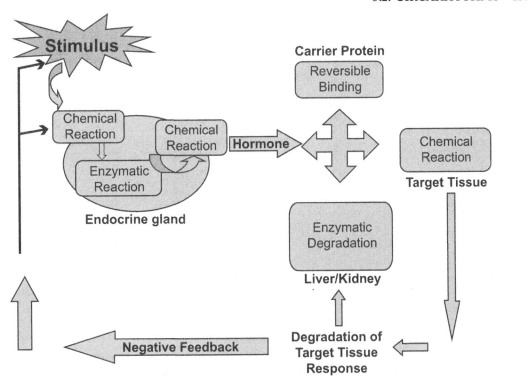

Figure 5.8: Model diagram displaying the flow of reactions taking place in order to transduce a stimulus into a target tissue response.

only stimulus we are certain affects the circadian oscillator. For the more mathematically inclined, Chapter 6 discusses the interactions between oscillators and external stimuli in a far more elegant fashion. See also discussions by Winfree[100], Pavlidis[63], Pikovsky and colleagues,[66] and Forger and Paydarfar[32].

Nothing is ever that simple, of course, and changing the model has interesting consequences. For example, if we revisit the simulations displayed in Figure 5.9 with the autorhythmic component now downstream from the endocrine gland, a different result is observed as shown in Figure 5.11. In this case, although the amplitudes of some components of the oscillating system are reduced, the system remains partially rhythmic. In this version of the model, carrier protein activity was made autorhythmic which buffers the system somewhat against the effects of the constant high intensity stimulus.

To be sure, these models are not an accurate reflection of the detailed molecular mechanism by which circadian oscillators work. The models are too simple to truly represent living systems that are incredibly complex. However, what these models do demonstrate is that feedback oscillators of this

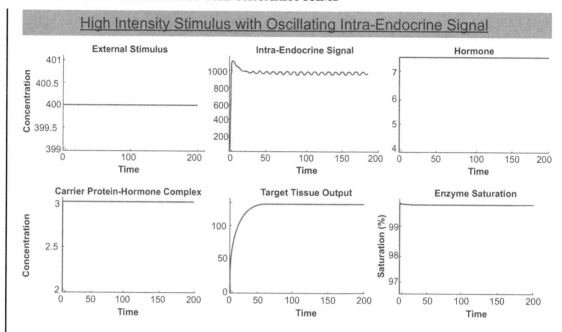

Figure 5.9: Endocrine-hormone system with an oscillating intra-endocrine signal that becomes damped out by a constant high intensity stimulus. (Modeled by K. Freedman.)

kind can be sensitive to the duration and intensity of input stimuli. The model provides what might be described in legal proceedings as "probable cause." They provide enough of a reason to launch a search for any evidence for or against the hypothesis that constant light of sufficient intensity can disrupt circadian rhythms. In fact, this hypothesis has been confirmed empirically[4,7,42]. Thus, we now have two potential methods for creating temporal disorder in circadian systems.

The third potential method also relies on input stimuli—light, as before—but in a far different way. So far, the models we have considered either use a short, intense burst to stop the rhythms or a high level of constant stimulation to drive the system to a constant level and away from oscillations. The final approach involves exposing the system to rapid changes in the timing of the LD cycle and relies on time delays within the circadian organization to disrupt overall temporal coherence.

This model works as follows. Suppose an LD cycle is used as the geophysical rhythm that synchronizes a circadian oscillator in a model similar to that displayed in Figure 5.1. The ultimate function of the oscillator is not to simply fluctuate in line with the LD cycle but rather to coordinate the rhythms downstream so that the entire organism's temporal structure is integrated into a coherent whole. However, looking closely at Figure 5.1, there has to be a time lag between changes in the oscillator as it responds to changes in the Zeitgeber and changes in biological rhythms as they respond to alterations in the oscillator. This is symbolized in the figure by the arrow labeled C_{pr}.

A

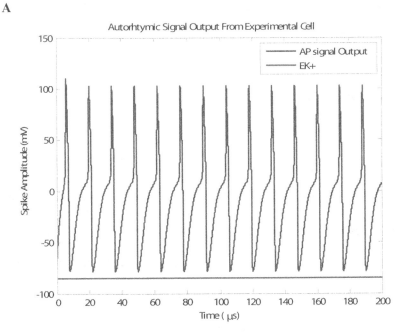

B

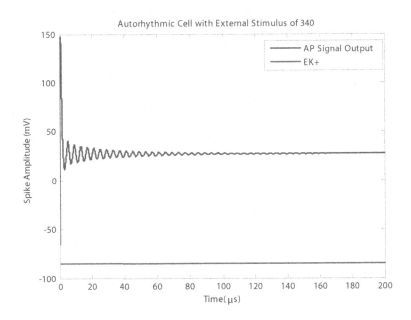

Figure 5.10: Autorhythmic neurons. (A) Without external stimulation; (B) Exposed to high intensity, constant stimulation. (Modeled by G. Neusch and R. Gangulty.)

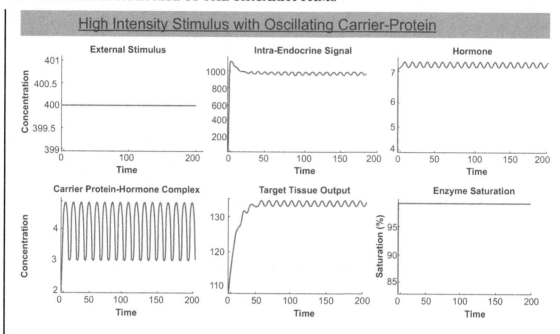

Figure 5.11: Endocrine-hormone system with an oscillating carrier-protein and a constant high intensity stimulus which does not damp out the rhythm. (Modeled by K. Freedman.)

This represents the coupling between the pacemaker P and the output rhythms, R. Even if the oscillator/pacemaker were able to respond instantly to changes in the timing of the LD cycle, it would certainly take time for the biological rhythms to readjust to the new positioning. The new timing information would have to be communicated in some fashion to the cycling physiological processes and these cycles would then have to be adjusted.

During the period when the biological rhythms are not fully resynchronized to the new chronological pattern as dictated by the new temporal position of the LD cycle, the organism can be thought to be *externally* desynchronized. Even more destructive to rhythmic coherence is the likelihood that the connections between the circadian pacemaker and various output rhythms will have different coupling and thus different time lags. In effect, many different physiological cycles will probably be tied to the pacemaker via arrows of different sizes (recall Figure 1.6 in Chapter 1). Under those circumstances, not all the physiological cycles will adjust to new temporal "orders" from the pacemaker at the same rate of speed. Thus, not only will the organism's daily biological rhythms be desynchronized from the external LD rhythm, but multiple internal cycles will be desynchronized from each other as well. This leads to the possibility of simultaneous *internal* and external desynchronization. That could almost serve as a definition of rhythm disruption.

The situation gets even more interesting as the circadian system becomes more complex in its organization. Consider the more evolutionarily realistic situation shown in Figure 5.4 where a single pacemaker is associated with several internal ("slave") oscillators. In this case, changes in the LD cycle are first communicated to the master pacemaker which, in turn, must resynchronize the secondary oscillators. The internal or secondary oscillators then must reestablish the proper temporal order in those physiological rhythms under their control. Since the time frames for these activities are going to be slightly different in each subcomponent of the system, constantly changing LD cycles are likely to generate considerable long-term external and internal desynchronization within any organism exposed to such unstable cycles (see Pittendrigh[67,68] for a discussion of the effects of LD shifts and slave oscillators). Experimental evidence from laboratory animals[9,72] and humans[6,34,48] supports this hypothesis that rapid changes in LD cycles can disrupt circadian rhythms.

In summary, there appear to be three basic methods for disrupting circadian rhythmicity in organisms:

1. The application of a light pulse of sufficient strength at a single point in the circadian cycle;

2. Exposure to a constant level of high intensity light;

3. Exposing the organism to constantly shifting LD cycles.

Of course, there is one additional method that can be deduced from the basic models displayed in Figures 5.1 and 5.4, which necessarily would be restricted to animal studies for *direct* observation. If the identity of the pacemakers were known, and if they could be damaged or destroyed without significant damage to the rest of the creature, then one could simply eliminate the pacemaker(s) and observe the results. Have any of these methods been applied and if so, are there data to indicate that circadian disruption generated any deleterious effects on the organism whose rhythms were thus disturbed?

5.3   HUMAN STUDIES

Human studies on the potential deleterious effects of circadian rhythm disruption suffer from the same limitations as were previously discussed concerning the attempts to associate smoking with lung disease. It is clearly unethical to deliberately subject human beings to stimuli thought to generate harmful effects in order to determine just how harmful those effects are. Thus, studies involving controlled experimentation on the effects of human rhythm disruption are limited. I know of no studies where the circadian rhythmicity of human subjects was purposely stopped or where humans were exposed to constant levels of high intensity light for prolonged periods under experimental conditions to determine the effects of such treatment on physiological or psychological well-being. Human subjects have been studied in temporal isolation, but with the object of learning about their circadian organization and not disrupting it. In fact, humans do have a marked tendency to develop internal desynchronization while exposed to temporal isolation, but for the most part, these observations have not been so prolonged as to generate significant physiological abnormalities[8,61,98].

Some studies have deliberately induced such desynchronization to study human circadian organization,[99] but again, the purpose of the studies was to investigate human rhythmic organization and not ascertain how disrupting rhythms affected normal function. Nonetheless, modern society does provide, as was the case with smoking, significant sources of data from which correlational studies can be drawn. The data come from the third of the three possible methods of generating circadian disruption, rapidly changing Zeitgeber cycles. The sources for these cycles are jet travel across multiple time zones and shift work.

5.3.1   TRAVEL ACROSS TIME ZONES—THE PHENOMENON OF "JET LAG"

Many of you may have traveled across multiple time zones and experienced the mild discomfort commonly known referred to as "jet lag." My own experience of this phenomenon could best be described as a sense of mild disorientation when arriving at my destination. I often feel as if I was operating in a light fog, almost as if there was a translucent cloth over my eyes. Things do not seem quite right even if nothing feels really wrong either. This feeling seems to get worse when traveling over several time zones, especially during trips eastward toward Europe (I live in the Philadelphia area of the Eastern United States). I did not seem to suffer quite as much going to Hawaii, although the destination might have had something to do with that result. Technically known as *desynchronosis syndrome*,[10] jet lag refers to the set of symptoms commonly experienced by travelers when crossing multiple time zones. These symptoms are primarily behavioral and include temporary changes in the organization and duration of sleep, fatigue, difficulty concentrating, and a general feeling of malaise. As I write this, later today, I will be traveling to Singapore from Philadelphia, a trip that results in the maximum shift in timing possible—12 hours. Reports from colleagues having made similar journeys indicates that the discomfort and disorientation associated with this level of time zone shift will be significant.

A survey of business travelers by Rogers and Reilly[76] reported that 74% of survey travelers reported some form of jet lag. Of these respondents, 50% reported above average tiredness and fatigue and 28% indicated some disruption in normal sleeping patterns. A smaller percentage–5%–reported difficulty in concentrating, while 5% reported eating problems. Other studies on jet lag report similar findings, including insomnia and sleep disruption[5,21,45,79] and decrements in cognitive performance[40,47,48,49,50] associated with jet plane travel. Interestingly, Sasaki and Endo's[79] study shows that while east-west travel across time zones generated altered sleep patterns, equivalent north-south travel did not, supporting the hypothesis that rhythm desynchrony was responsible for the results rather than simply the stress associated with flying (see Carruthers and colleagues[12]). The prevalence and predictability of these symptoms led to the development of a questionnaire—the Columbia Jet Lag Scale[82]—to reliably measure the phenomenon.

Not all symptoms associated with jet lag are so innocuous, however. Tec[90] reported on the case of a man whose clinical depression was reliably precipitated by plane travel across 4–5 times zones. Similarly, Katz and coworkers[45,46] reported that a mild depression could be associated with jet lag and argued from data collected during a six-year study that travel across several time zones exacerbates major psychiatric disorders in susceptible individuals. Another interesting phenomenon associated with repeated jet travel involved altered menstrual cycles in airline personnel. Preston and colleagues[70] reported that 8/29 stewardesses surveyed reported menstrual irregularity, while Iglesias and coworkers[43] found that 48% of 200 stewardesses surveyed reported some form of significant change in their menstrual cycling associated with travel across time zones.

Unfortunately, studies involving actual air travel confound the effect of time shifts with the discomfort and stress of flying. For those of you who have flown recently, I am sure you know what I mean. Sometimes, it appears that airlines actually go out of their way to design the most uncomfortable seating arrangements possible. After 5–7 hours of sitting in a seat clearly designed for someone half the bulk of the average human being, with your knees embedded in the seat in front of you, listening to endless inane chatter and while breathing dehydrated, recycled air at varying pressures, it is not surprising that the average air traveler is a bit irritable—to say the least.

However, since most individuals appear to suffer from fairly mild symptoms and given that people generally recover within 2–3 days from the magnitude of phase shift associated with jet lag[78], it is ethically possible to test the idea that it is the phase shift creating jet lag's symptoms and not the act of flying. Taub and Berger[86,87] exposed volunteers to 2- and 4-hour phase shifts—both delays and advances—to determine the effects of such changes on sleep and performance. Not only were the expected sleep changes observed and a mild decrease in cognitive performance, but significant changes were also reported in cheerfulness, the ability to concentrate, and reaction time after individuals were exposed to the phase changes. Preston and colleagues[70] also ran tests in the laboratory and saw decreases in performance with phase shifts. Thus, it appears that at least some of the symptoms associated with jet lag really are due to the time shifts and the circadian desynchronization that results from those shifts.

5.3.2   THE 24/7 SOCIETY—SHIFT WORK, CIRCADIAN DISRUPTION, AND HEALTH ISSUES

Another relevant feature of our modern industrial society is the prevalence of shift work. In agrarian cultures, work follows the sun, beginning with sunup and ending with sundown. With electric lighting and modern engineering technology, there are no temporal limits to the ability to work, and activity can proceed across the entire 24-hour day. While machinery may be able to work continuously without regard for time-of-day, individual human beings cannot, and thus it becomes necessary to create shifts so that all parts of the day are adequately covered. The results are the three shifts typical in industrial societies: day shift (8am–4pm); swing shift (4pm–12 midnight), and night shift (12 midnight–8am). The actual times vary a bit from job to job, and overlaps are often built in the schedule to make certain the transition from shift to shift is a smooth one.

These working shift schedules do not, in and of themselves, imply circadian disruption by any of the three criteria discussed above. If an individual works consistently in one shift, and if that schedule is imposed during days off as well, then there would be no *a priori* reason to suppose that an individual might not simply entrain to whatever cycle they follow without significant desynchrony.

Of course, this does not normally happen. Despite the 24-hour-a-day character of modern society, a significant portion of societal functions are concentrated during daylight hours. The majority of people still work some form of a day shift, schools generally operate during the day, and most family functions are concentrated such that a permanent assignment to either swing or night shift could be highly disruptive to an individual's social obligations.

As a consequence, the rotating shift cycle was developed to fairly partition working in the various shifts across the employees. Under this strategy, workers rotate between each shift over a period of several days and never permanently assigned to a particular shift. This generates a classic paradigm of circadian disruption by continually altering the phases of the environmental cycles such that complete external synchronization is not possible. As discussed above, such a rotating schedule has a very high potential for generating internal as well as external desynchronization, resulting in workers suffering from a fairly high degree of circadian rhythm disruption. Even those workers who habitually work on swing or night shift do not maintain that schedule on their days off and thus will also suffer from some degree of desynchrony by switching from day shift (days off) back-and-forth to either swing or night shift when working.

Thus, there are actually four basic "shifts" involved in shift work: day shift, swing shift, night shift, and rotating shift. In addition, there are subcategories of rotating shift depending on the speed and direction of the rotation. For example, a worker could rotate from day to swing to night every two days or every week or every month depending on the circumstances. These different patterns will result in different degrees of disruption. Also, a worker could shift from swing to day to night shift or night to day to swing shift and these also might vary in their effects. All of this means that one has to be rather careful and precise in describing the actual operating procedures before associating any health problems with being a "shift worker."

Finally, we must always keep in mind the existence of confounding effects. One summer, I was hired to work in a plastics factory. The business specialized in a few basic products—plastic cement pans, window coves, and pallets—as well as recycling plastic by melting it down and extruding in long sheets. These sheets were then heated and formed into the products using large vacuum-generating machines that rotated and generated a new product every minute or so. Each time the machine rotated, we had to take off the formed product and cut the trim off semi-molten forms using long box cutters. It was hot, exhausting, and mildly dangerous. The factory was also located in Denver, Colorado, and I was assigned to the night shift.

Now, researchers often contend—and for good reason—that the night shift generates more accidents and less productivity than the day shift. The reasons are associated, as we might expect, with circadian disruption and sleep deprivation. So why did my night shift group generate fewer accidents and more productivity than either the day or swing shift? We were not super heroes and presumably as desynchronized and sleep deprived as night shift workers elsewhere. The answer probably lies in the nature of the work, the season, and the factory's construction. This factory had no air conditioning on the floor and, combined with the heat generated by the operating machines and melting plastic, the temperature within the factory during the day in the summer could reach

110°F or more. On the other hand, Denver's heat is a "dry heat" with low humidity, and it cools off significantly at night. So, the factory was seldom warmer than 85°F at night.

This had two important consequences. First, it was far easier to work longer and harder at night compared to the day. Second, many of the better workers wanted to work the night shift in the summer to avoid the excessive daytime heat. If a researcher were to have used us as a sample to study the effects of circadian desynchrony associated with night shift work, he or she would have concluded that circadian desynchrony has positive effects on job performance! Most likely this would be an incorrect conclusion, because the real-world confounding effects of factory conditions had not been adequately considered.

These confounding effects are numerous and difficult to control. On the one hand, some factors tend to decrease the reported negative effects of shift work on human behavior and physiology. Some individuals, finding themselves unable to cope with shift work, may elect to transfer, if possible, to day work. This leads to the possibility of self-selection, meaning that the population of shift workers is made up of a special population more tolerant to circadian disruption than would result from a simple random sample. There is also some evidence of under-reporting of health issues among shift workers who assume that such problems simply "come with the territory" and are thus not worth bringing to the attention of medical personnel[60]. On the other hand, recruitment of individuals of a specific socioeconomic status or type of work at greater health risk or association of a specific risk factor, such as smoking, with shift work could artificially inflate the perceived health hazards associated with shift work[101]. Finally, there are a number of methodological issues and concerns with both data collection and analysis.

Many studies are based upon questionnaires and surveys which are known to suffer from both volunteer bias and reliability issues. Shift work is also classified in many different ways, making comparisons across various studies difficult. Finally, shift work is often used as a "catch-all" term for any type of work schedule that deviates from standard daytime work hours. This confounds the effects of extended work hours (e.g., 12 hour shifts), rotating shifts and schedules which are based upon fairly stable evening and night shift work times. These schedules differ in both their effects on circadian rhythms and in the types and extent of confounding factors associated with them, making reliable conclusions problematic.

Keeping these limitations in mind, what do the available data indicate about the effects of shift work on human health? The majority of studies report negative effects associated with shift work. These effects can be broadly classified as behavioral problems, abnormal or disrupted sleep[2,3,31,53,96,97], reproductive issues[25,83], gastrointestinal complaints[13,25,65,85,94], metabolic abnormalities[35], increased risk for cardiovascular disorders[54,84,92] and an enhanced susceptibility to certain cancers[17,18,37,80,81]. Behavioral issues range from decrements in performance[11,38] to depressed mood[41] to increases in neuropsychological problems[1,14,52]. Numerous reviews have been published linking shift work with various health problems[11,15,38,39,53].

Most is not all, however, and there are a number of reports that do not support the hypothesis that shift work increases risk for health problems. Fuijta and colleagues[33] found no association between shift work and any form of illness, while Taylor and coworkers[88,89] reported equivalent suicide rates between day and irregular shifts and lower incidents of psychoneurotic problems with shift workers as compared to day workers. Although Sveinsdotter[85] did find an association between rotating shift workers and gastrointestinal complaints, he found no effect on job satisfaction, the incidents of illnesses or sleep quality associated with shift work on a self-assessment survey. In Costa's[15] review, he lists 25 studies from 1939 to 1987 involving over 68,000 subjects which reported that shift work increased gastrointestinal problems, 10 studies from 1951 to 1987 involving over 29,000 subjects which could not find such as association, and even one study from 1963 involving 561 individuals which reported that day shift had more problems!

For anyone aware of the years of study and controversy surrounding the investigations linking smoking with lung disease, there should be a familiar sense of *deja vu*. You have seen this before and for much the same reason. Correlational studies simply cannot establish causality—for that, we need controlled experiments. No matter how many confounding effects one attempts to control for in a correlational approach—and many researchers have made heroic attempts to do so—there is always that possible effect that was not controlled because the researcher was simply unaware of its importance. In the case of smoking, the variety of correlational studies eventually made it highly improbable that some unknown confounding effect was hiding in the wings, waiting to be discovered. The same argument could be made in this case but, fortunately, it is not quite as necessary. Unlike the study of smoking-related health issues though, where confirmation with animal experiments has been extremely difficult to achieve, some significant experiments examining circadian disruption have already been conducted with animals. The results of these experiments support the proposition that circadian disruption has deleterious effects on health.

5.4   ANIMAL STUDIES

The subjects for these experiments have primarily been laboratory mice, rats, and hamsters subjected to a variety of conditions designed to generate circadian disruption. The health effects monitored in these experiments include gastrointestinal problems, cardiovascular disorders, certain cancers, and learning. The treatments used to create circadian disruption included constant light (LL), repeated phase shifting of the LD cycle, and exposure to abnormal LD cycles to which normal synchronization was not observed. In addition, some experiments involved elimination of a known master pacemaker or circadian clock (the suprachaismatic nucleus), while another examined the effects of a genetic mutation known to alter the organism's circadian rhythm structure. Since the purpose of this chapter is to provide a proof-of-concept for the impact of rhythms on health and performance, I will review these experimental data based upon the impact on specific physiological and behavioral functions.

Insofar as gastrointestinal problems are a commonly reported problem associated with shift work, let's begin with experiments confirming the link between disrupted circadian rhythms and GI abnormalities. In 1994, Larsen and coworkers subjected rats to LL for 4 weeks. Measurement of gastric rhythms revealed changes in pepsin secretion and potential differences across the endometrial lining. When examined, 69.8% of rats subjected to LL had GI lesion scores exceeding 1.0, while none of the rats maintained in a standard LD 12/12 cycle had a score greater than 0.5. More recently, Preuss and colleagues[71] examined the effects of repeated phase shifts (12 hours every 5 days for 3 months) in mice. Using a model of dextran sodium sulfate-induced colitis, the results showed that phase shifting treated mice accelerated the progress of the disease, resulting in reduced body weights, increased abnormal histopathology, and an increased inflammatory response compared with treated mice kept in a standard non-shifting LD cycle. Thus, it appears that repeated external and internal desynchronization disrupts normal GI function.

Similar results have been reported for the effects of rhythm disruption on cardiovascular function. Penev and coworkers[64] subjected Syrian hamsters carrying a gene for cardiac hypertropy to 12-hour phase shifts every 7 days. Compared with non-shifted mice, the animals subjected to repeated phase changes displayed an 11% reduction in longevity associated with cardiomyopathy. Using a mouse model of pressure overload hypertropy, Martino and colleagues[57] demonstrated that treated mice subjected to a 20-hour LD cycle displayed significant cardiac abnormalities compared with similarly treated animals kept in an LD 12/12 or 24-hour cycle. Even more significantly, mice subjected to the 20-hour cycle for 8 weeks and then returned to a normal 24-hour cycle displayed progressive reversal of cardiac symptoms with the return to internal and external synchrony. In an earlier report, Martino and coworkers used the so-called *tau* hamster mutation to investigate the effects of desynchrony on cardiovascular function. The tau mutation is an autosomal recessive which results in a shortening of the inherent circadian period of affected hamsters in DD. Normal wild-type (+/+) hamsters have a period of nearly 24 hours in DD, heterozygotes (+/tau) display a 22-hour cycle, while homozygotes (tau/tau) show a 20-hour DD cycle. As it turns out, heterozygotes suffer from a reduced lifespan, apparently associated with cardiovascular and renal dysfunction. Interestingly, this effect is seen only when the animals are subjected to a 24 LD cycle, resulting in internal and external desynchronization. When the animals are raised under a 22-hour LD cycle, the pathology does not occur, and the lifespan is normal.

In summary, experimental evidence indicates that circadian disruption is a risk factor for cardiovascular pathology, whether that disruption is induced through repeated phase shifts or exposure to abnormal LD cycles to which the organism cannot adequately synchronize.

A number of studies have demonstrated a link between circadian disruption and cancer. One of the earlier studies involved the use of diethylnitrosamine as an agent to induce liver carcinogenesis in rats. The study compared two methods for enhancing the drug's effects—phenobarbital and exposure to constant light (LL). The results indicated that LL was as effective as phenobarbital in generating liver cancer in this model and both were much better than simply maintaining the animals under a standard 12/12 LD cycle[93]. In 2004, Filipski and colleagues[26] used a phase shifting approach, exposing mice to 8-hour phase advances every 2 days. After 10 days of exposure to either shifting or stable 12 LD cycles, mice were inoculated with Glascow osteosarcoma and examined 15 days afterward. The examination revealed that exposing the animals to phase shifting conditions significantly accelerated tumor growth[26]. This experiment was repeated[27,28] to ensure the causal nature of the effects. Interestingly, in one case, mice were subjected to a meal timing protocol which is known to help stabilize rodent rhythms. This procedure partially counter-acted the effects of the repeated LD cycle shifts[27]. Finally, Filipski and coworkers subjected mice to ablation (destruction) of the master circadian pacemaker—the SCN—and inoculated these and sham-operated animals with both Glascow osteosarcoma and pancreatic adenocarcinoma. SCN-lesioned animals displayed tumor growth rates between 200% and 300% faster than their sham-operated counterparts[29]. Figure 5.12 shows the effects of SCN lesions on these mice, while Figure 5.13 displays the effects of SCN ablation—and thus circadian disruption—on tumor growth[28].

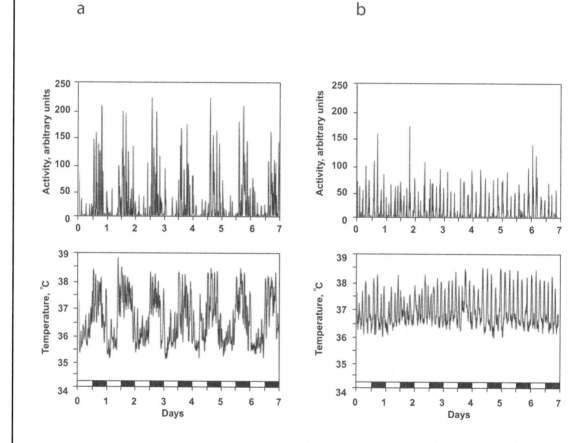

Figure 5.12: Activity (top graphs) and temperature (bottom graphs) recorded from a sham-operated mouse (left) and a mouse with a SCN lesion (right). (From Filipski, et al., 2003, with permission.)

Experiments by Craig and McDonald[16] examined the effects of phase shifting on learning and behavior in rats. In this paradigm, rats would be subjected to a 3-hour-per-day phase advance for 6 days followed by a recovery period of 10 days' exposure to a stable LD cycle. The studies were divided into *acute* and *chronic* experiments. Acute experiments consisted of a single 16-day protocol,

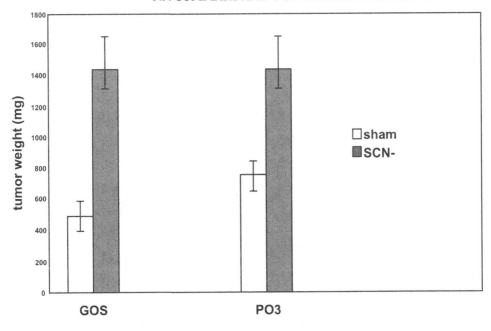

Figure 5.13: Differences in tumor weights between sham-operated normal mice (clear) and mice with SCN lesions (grey). (From Filipski, et al., 2003, with permission.)

while chronic experiments exposed the rats to four such protocols lasting a total of 64 days. The results indicated that while acute experiments had no significant effect on learning and behavior, chronic experiments generated a significant deficit in the rat's ability in a water maze task involving spatial memory[16].

Finally, the adaptive advantage of intact circadian rhythms was examined in experiments by DeCoursey and colleagues[19,20]. In these experiments, the behaviors of white-tailed antelope, ground squirrels, and eastern chipmunks were compared for the effects of disruption in circadian rhythms through ablation of the SCN—circadian pacemakers. The ground squirrels (Figure 5.14a) were observed in a semi-natural environment under predation pressure by weasels. The chipmunks were caught, treated, and released back into a natural environment (Figure 5.14b). As can be seen in Figure 5.14, mortality was significantly greater in those animals without coherent circadian cycles.

5.5   SUMMARY AND CONCLUSION FOR PROPOSITION 2.3

The evidence indicates, in both humans and animals, and from both correlational and experimental studies, that disruption of daily rhythms is a significant risk factor for certain physiological and possibly behavioral disorders. Although the exact mechanisms are not clear, and indeed may differ

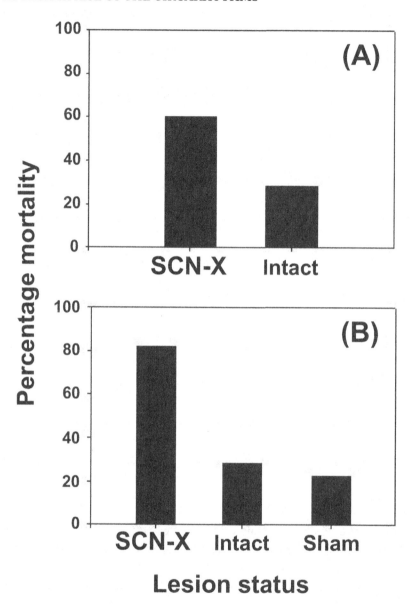

Figure 5.14: Survival of ground squirrels and chipmunks when the suprachiasmatic nuclei (SCN-Circadian Pacemakers) are lesioned, causing a loss of coherent circadian rhythmicity. (From Paranipe and Sharma, 2005, open source.) The top graph (A) shows the effects of SCN lesions on ground squirrel survival in a semi-natural environment, while the lower graph (B) shows similar results for chipmunks in a natural environment.

in each case, the available evidence strongly supports the validity of the proposition that circadian rhythms are important to maintaining health and well-being.

5.5.1   MEASURING TIME'S PASSING—THE ADAPTIVE FUNCTION OF INTERVAL TIMERS

We now know that oscillations can be generated through various feedback systems. In Chapter 7, we will discover that oscillations may even evolve as a derivative characteristic of as fundamental a process as neural signaling. As a result of our investigations, we have concluded that autorhythmic or endogenous oscillators process several key elements that provide for stable internal and external synchronization. Thus, we can anticipate how these biological clocks might have evolved and are not surprised we find such devices in circarhythm systems.

From an evolutionary standpoint, these oscillator-based systems provided two basic adaptations. First, oscillators can provide for the internal temporal organization of activities needed for any complex, goal-oriented device. Second, these systems can fit internal activities within a temporal ecological niche so that there is a match between an organism's internal activities and what is happening—or about to happen—in the external environment. Coordinating all these activities is what we have referred to as maintaining internal and external synchrony.

So, to quote a song from 1969 sung by Peggy Lee, is that all there is? If you consider how you use timing devices in your daily routine, you are forced to conclude the answer is no, there is more to it. Returning to my kitchen where our journey began in Chapter 1, there is a timing device that I failed to mention. How is that possible? I referred to a clock on the stove, one on the microwave, and even one linked to the United States Atomic clock hanging on the wall. No wonder I am writing about biological rhythms—if I have yet another clock, I must have some kind of time fetish. However, I would hazard a guess that you also have this alternative timing device in your kitchen as well. It may not be obvious at first because it is not running all the time, like a clock. It only works when specifically set. As I am sure you recognize by now, this timing device is the food timer built in to both the microwave and the stove.

Why are there two kinds of timers? Wouldn't one be sufficient? To answer this, consider how you use these two timers. At the moment, the timer on the microwave is set for 2 minutes for my tea, while the oven timer has about 12 minutes left before I go out to change the watering on my lawn. Now, I could have simply used a clock for both activities. I could have noted the current time and set an alarm to go off after two minutes for my tea and two hours for the watering system. It would have worked fine, although the temporal resolution might have been a problem for the tea. Two minutes is a bit short for an alarm clock and is difficult to set. In addition, most clocks do not allow you to set alarms in second resolution, and thus exactly 2 minutes might be bit difficult to arrange. However, theoretically at least, it could be done. So why do we use this other kind of timer—an *interval timer*—in addition to a clock?

The fact is that in these latter cases, the local time is not important. We are not concerned if it is 4:06pm on a Sunday afternoon or 7am Monday morning when determining the length of time for generating our tea. Two minutes is two minutes and when the interval is the only important factor, an interval timer is an efficient and effective method of measuring time. Why, then, develop clocks at all? Could interval timers not do the job equally well?

Again, if you consider this issue in terms of your day-to-day experiences, you can see how an interval timer might be used as say, an alarm clock. Each night, you could determine when you wanted the alarm to go off and you could set the number of hours and minutes accordingly. However, for most of us, we are not as concerned with the time spent asleep as we are with the time we need to wake up. In this case, local time is important. Most of us need to get to work or school at a specific clock time, say 8am. Using an interval timer, we would need to get the current time when we went to bed (necessitating some form of clock regardless), calculate the amount of time between the current time and the wake up time, and then set the alarm. Assume that we need an hour to get up and get to work. Then, we would need to wake up at 7am. If we went to bed at midnight, the timer would be set for 7 hours and if we stayed up late to watch a movie until 2am, the time would need to be set to 5 hours. Every night, we would have to recalculate the timing and reset the timer. If we forgot to set the timer, we might well be late to work. How much easier it is to set an alarm clock for 7am and then not worry about it any further. Because the clock is a self-sustained oscillator, all we need do to supply power, set the clock properly, engage the alarm, and the buzzer will go off at 7am every day. This might be a bit annoying on a Sunday morning, but then again, nothing is perfect.

We know from our previous discussions that biological systems incorporate oscillators and clocks into their temporal organizations. What about interval timers? To what purpose could such devices be put in living systems? To understand this, let us examine two extreme examples of timing for a specific behavior. In our case, let's use foraging behavior in which an animal searches for food.

Let us begin with the more familiar oscillator functioning as a biological clock. In Figure 5.15, the basic components are displayed. The function of the oscillator is to correctly position the behavior, in this case foraging, in the ecologically appropriate time period. Since this time period is displayed as night, we know that we are dealing with a nocturnal animal. For various evolutionary reasons—predator avoidance, food availability, etc.—the species have evolved so as to concentrate activity in darkness. The oscillator creates the appropriate timing by initiating and halting foraging activity through some kind of internal signal, indicated by the green threshold line, which is synchronized to the geophysical cycle of night and day. The oscillator could accomplish this by first entraining to the day/night cycle via a phase response curve. Then, as the oscillation rises above threshold, signals could be sent to awaken the animal and increase its sense of hunger. A hungry, awake animal is then likely to engage in foraging behavior. As the oscillation descends below the threshold line, the animal could then be signaled for sleep and hunger diminished, rendering the animal inactive until the next wake up call.

So far, so good—this is the reasoning that we have been following all along. It makes sense that such activities would be temporally organized from *biochemistry*, such as synthesis of digestive enzymes in preparation for food consumption and liver enzymes for metabolic regulation and biotransformation, to *physiology*, as in the secretion of those enzymes as well as other digestive pro-

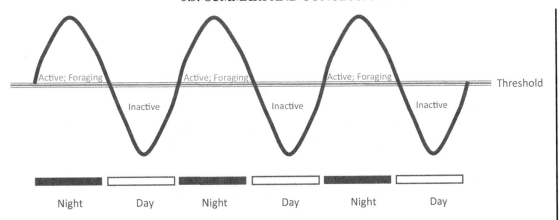

Figure 5.15: Oscillator control of foraging behavior.

cesses, to *behavior* resulting in the active foraging and consumption of food items. All synchronized, efficient, prepared, and pre-positioned. Perfect, except for one potential problem. There is nothing in this system to account for the actual level of food consumed (or not consumed). What happens if the foraging for one night is unsuccessful?

Interestingly, the answer in this oscillator-only model is nothing. There is no change in the system. As the oscillation decreases below threshold, hunger decreases, the propensity for sleep increases and the animal rests until the next rise above threshold. Success or failure in terms of foraging is simply not important to the model's functioning. Success or failure may not be important to the model, but is quite likely to be *very* important to a living animal.

How important this is depends on the specific animal and the environmental circumstances to which the species have become adapted. If, for example, there is a very high probability of death if the animal becomes active during the day, most likely the species will evolve a capacity to fast long enough to wait until the next cycle to eat. Evolving an automatic appetite suppressor associated with passing below the threshold of a synchronized oscillator would accomplish that adaptive goal. However, no matter how dangerous daylight might be for the species, at some point, animals must eat to survive. Simply suppressing appetite will not overcome lack of food forever. If food is available only during daylight, at some point the animal must become opportunistic enough to take advantage or starve to death. Unfortunately, a simple oscillator model synchronized to the day/night cycle does not appear to have the flexibility to adapt to such an unusual circumstance.

How would an interval timer be advantageous in such a situation? Recall that interval timers are not necessarily tied to the actual time-of-day but rather to the passage of time. That is what we used our microwave and stove timers for in the examples given above. As we did with clock time, however, to understand what is meant by passage of time for non-human organisms, we must translate from our technological perceptions to the perspectives of a living system. Organisms tend to be affected by relative intervals of time rather than the exact number of hours, minutes, and seconds. Just as an absolute time, such as 10am, is not the issue for a bird *per se,* but rather the hours after sunrise when worms are available, it is not necessarily 6 hours that is meaningful for our nocturnal animal but rather the time since the animal's last meal. The advantage of an interval timer in this case is that it can be more directly tied into the animal's metabolic requirements when compared to an oscillator. For this mechanism to evolve, the animal's survival needs must be sufficiently great to occasionally override the selective advantage of being inactive during a particular time of day. Otherwise, there is no particular adaptive reason for not relying on the oscillator-based models already discussed.

Consider the interval timing system presented in Figure 5.16. When an animal passes below the threshold for hunger, a trigger for active foraging is tripped and the animal begins foraging. Foraging creates a buildup of fuel (indicated by the red line) in the animal that continues until a satiety threshold is reached. At that point, the animal stops foraging and the available internal fuel begins to decrease until the hunger threshold is reached and the process begins again. Note that there is no time-of-day data included in Figure 5.16 since the model does not require it.

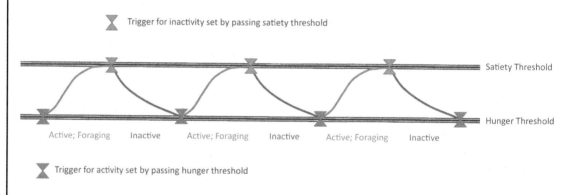

Figure 5.16: Model of an interval timer controlling foraging behavior.

In Figure 5.17, the combined effects of a circadian oscillator tied to the light/dark cycle and a metabolic interval timer are displayed. In the figure, the red lines are when the interval timer would generate foraging activity while the blue lines are the inactive phases of that same interval timer. For the circadian oscillator, activity and foraging occurs when the oscillation rises above the threshold while inactivity and sleep are promoted below the threshold. Although the two rhythmic mechanisms produce rhythms that are slightly out of phase, it is clear that the combined effects of these two timing systems create a coherent temporal order.

In that case, why have two different systems? Consider what happens if an animal cannot find enough food during a normal nocturnal foraging period. This situation is diagrammed in Figure 5.18.

In Figure 5.18, the animal fails to consume enough food to reach a point of satiety. In this extreme example, it appears that the animal does not locate any food at all. Thus, there is no build

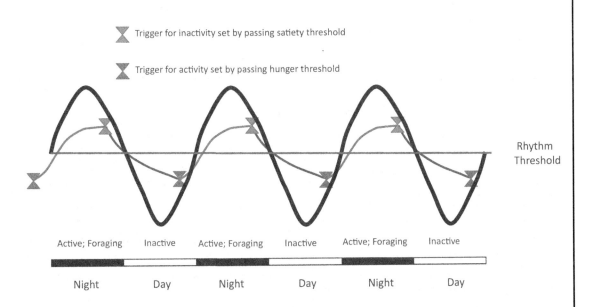

Figure 5.17: The combined influence of a circadian oscillator and interval timer on foraging behavior.

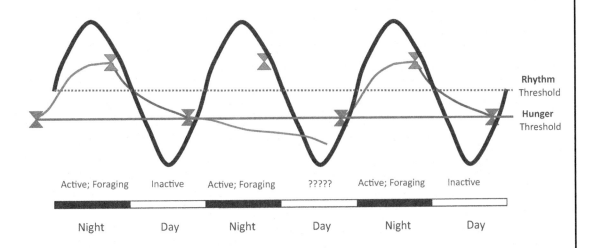

Figure 5.18: Diagram of combined interval timer and circadian oscillator during which the animal fails to find/consume enough food. During the second night, the animal cannot forage enough to reach the satiety trigger (upper brown hour-glass). Hunger thus increases throughout the night and into the next day.

up of the physiological signal indicating that food is being consumed and the red line continues downward. This downward trajectory is interpreted by the animal as an ever-increasing level of hunger that extends into the daylight hours of the next cycle. This sets up a conflict between the interval timer and circadian oscillator. The interval timer is now promoting active foraging behavior using hunger as the stimulus. At the same time, the circadian oscillator is promoting inactivity and rest, possibly through appetite suppression.

So what is the end result? Which timer wins? As usual, the answer is that it depends. More specifically, it depends on the balance of factors in terms of the costs and benefits of fasting vs. daylight behavior. These factors will differ between species, individuals, and environmental circumstances, so a single answer is simply not possible. Organisms position their behavior within the geophysical cycle so as to maximize their genetic propagation (Evolution 101 again). Here, it means that there must be a cost/benefit ratio for activity at different times of night and day which has led to the temporal positioning (temporal niche) at night. At the same time, animals cannot survive without fuel in the form of food. The conflict can be represented in the graph displayed in Figure 5.19.

This is a model of a real conflict that might arise between maintaining temporal order and synchrony and the lack of a necessary resource—food in this example—at the usual time. Of course, the numbers and shape of both curves would need to be experimentally determined for a specific species or organism. But even though it is a schematic, the graph suggests how such conflicts could be resolved. On the left side of the curve, the costs of temporal disruption (or the benefits of temporal order) exceed those associated with a temporary fast. Thus, in that range the circadian oscillator "wins" and will dominate the activity cycle of the animal, even though the animal may remain somewhat hungry during the inactive phase of its cycle. However, as time goes on, the cost of fasting steadily increases. At the transition point, the costs of fasting and those of temporal disorder coincide. If food does not become available at this time, the costs of continued fasting will exceed those of temporal disruption and the animal will begin seeking food regardless of time-of-day. We would not expect this to happen very often—proper positioning of behaviors is one of the main adaptive advantages of synchronizing to geophysical cycles—but it is certainly conceivable. If the animal's entrained (to the day/night cycle) circadian oscillator were the only factor determining the time of foraging behavior, it is possible during times of food scarcity, the animal would miss opportunities for food gathering during daylight. Thus, the paradoxical result would be observed that the animal retained perfect temporal synchrony as it starved to death. Such an outcome does not appear particularly adaptive.

This result emphasizes a point that has been made previously but needs to be reiterated constantly when discussing biological organisms. There is a difference between the ultimate function of living systems—the preservation and propagation of genetic material—and the proximate mechanisms by which these ultimate ends are achieved. True, biological rhythms are a powerful and ubiquitous proximate mechanism subserving the ultimate function of organisms on Earth. However, the generation and maintenance of rhythms are not themselves fundamental objectives of living things; they are merely specific means to an end. If a conflict arises between rhythms and genetic propagation, evolutionary processes should, if possible, promote propagation at the expense

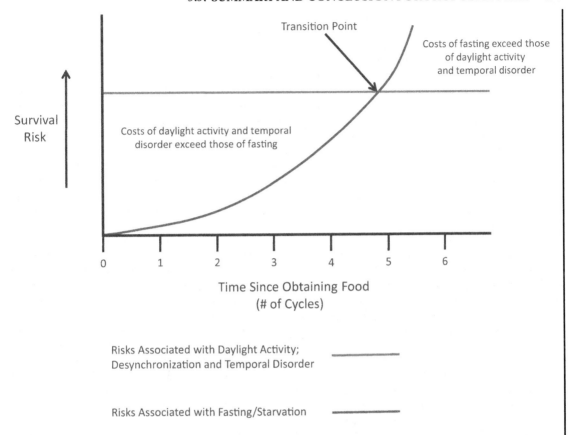

Figure 5.19: Graphical representation of the conflict between costs and benefits of maintaining stable synchrony with the geophysical cycle in the face of inadequate food availability.

of rhythms. In the above instance, by including an interval timer keyed to levels of metabolism associated with food intake, evolution has provided a kind of "fail-safe" to prevent the circadian oscillator from damaging the long-term survival prospects of an animal simply to maintain temporal order.

Some of you may be puzzled by the model's use of the term "temporal disorder." It is certainly understandable that model should consider the costs of being active during the day, and this shift of activity will cause some desynchronization, at least externally, between the internal circadian oscillator and the external day/night cycle. Why, however, was temporal disorder included as a potential cost in the model? Is this not an internal function and, if so, why would foraging during the daylight hours have any impact on it?

Temporal order is most certainly an internal function, and to understand how the alteration in foraging behavior might impact it, we need to recall the rhythmic cascade effects discussed in previous chapters. The circadian oscillator does not just affect foraging behavior alone, it orches-trates an entire symphony of biochemical and physiological activities and responses to optimize the metabolic utilization of food (recall our discussion of the "on-time" bird). Digestive enzymes must be synthesized, liver biotransformation systems brought on line, and countless other adjustments need to be made if the food is to be utilized most effectively. And just to make things a bit more complicated, the actual organizational structure of a circadian clock system could be quite complex. This complexity may actually enhance the likelihood of internal desynchronization in response to unusual events, such as the conflict with an interval timer.

Since circadian clocks are an evolutionarily ancient development, all eukaryotic cells most likely have an ancestral ability to generate self-sustained circadian rhythms as part of their joint evolutionary legacy. Using the heart as a model system again, we can imagine the evolution of the circadian system as the coalescence of the main clock functions into a few "master" pacemakers entrained by geophysical cycles which, in turn, entrain a multitude of internal (secondary) circadian oscillators. The system works well as long as there are strong, consistent signals between the geophysical cycles and the master pacemakers and between the master pacemakers and the internal, secondary oscillators. Any factor that weakened the coupling between the various components of the system would increase the likelihood that the system would become temporally disordered.

Enter the effects of the interval timer and the shift of activity into the daylight. This shift to daylight activity would have numerous effects on the internal rhythmic structure. It would allow light to impact the animal at unusual times, with unknown effects on the circadian oscillator's phase response curve. It would create changes in metabolic factors (assuming the animal was able to obtain food during the day) at odd points in the animal's cycle, again with unknown effects on the master and secondary oscillators. In fact, the change in timing would create numerous potential entraining signals with abnormal patterns which, as will be discussed in Chapter 6, could cause chaotic responses in the oscillators. In fact, given that rhythms cascade within complex systems, alterations in rhythms would propagate throughout those systems just as much. Internal temporal disorder is indeed a factor to be considered when changing the activity pattern of an organism, and such changes are not to be entered into lightly or without a clear adaptive advantage.

In summary, interval timers measure the passage of time and can theoretically do so without any reference to external geophysical time. There are a number of proximate mechanisms that could be hypothesized to create such timers, the simplest of which measure either the accumulation or loss of some substance. The substance could itself be some physiologically important material, such as serum glucose, or it could be a marker related to a significant chemical or activity. In this case, time is a quite flexible and relative concept, since it is the rate of accumulation or loss of the substance which matters and not the absolute physical notion of seconds and minutes. It is the role of the circa-oscillators to adjust the interval timers so that, under normal circumstances, they adaptively fit into significant geophysical cycles.

5.6   TIMING IS (ALMOST) EVERYTHING

Circa-rhythms, in particular, are subject to another set of confounding factors relating to the adaptive role of these rhythms in positioning biochemical, physiological, and behavioral activity into the proper temporal niches associated with geophysical cycles. When comparing circa-rhythms to other biological cycles, such as the cardiac cycle, a significant difference appears in the nature of the rhythms' function in the evolution of the species. While all biological cycles ultimately function to promote genetic propagation, the role of the heart as a pump is inherently rhythmic—it is simply not possible to separate the function of the heart from the cardiac *cycle*. The role of circa-rhythms, however, is to increase the effectiveness of various activities through temporal coordination. It is quite possible to separate the role of the rhythm from the role of the activity being coordinated. As an example, cortisol acts to increase serum glucose by promoting gluconeogenesis—the synthesis of glucose from, among other substances, certain amino acids. Now, we know from previous chapters that the levels of cortisol vary in human serum by some 250% over the 24-hour day. We also know that the substrates and enzymes with which cortisol interacts have their own substantial rhythms, and these cycles will cascade throughout the whole system, generating substantial differences across the 24-hour day. However, that result is not inherent in the description of cortisol's biochemical activity in the same manner as the cardiac cycle is inherent to pumping blood. A pump is inherently oscillatory—we cannot describe how a pump works without including the rhythmicity of the pumping action. But we can describe a biochemical reaction, in terms of its immediate function, as the linear consumption of input and production of output—even if a such a description would be inaccurate in full temporal context. This separation of rhythm from function generates both conceptual and empirical problems.

Conceptually, the separation allows people to ignore a vital part of how all living systems operate. It permits biology texts to barely mention circa-rhythms or ignore them entirely. It allows medical schools to train physicians to be ignorant of a fundamental characteristic of human physiology and behavior and lets workers be scheduled in a manner detrimental to their health and job performance. It supports a societal environment that generates untold suffering and loss. Ignorance, in this case, is not bliss.

Empirically, the role of circa-rhythms as temporal coordinators is a supporting, rather than primary role, for the scheduled activities. As such, the effects of a circa-clock may be altered or even partially overcome should circumstances warrant it. The use of interval timers, as discussed above, is one method by which evolution has "hedged its bets." Too strong an adherence to the dictates of a circa-rhythm can be maladaptive, and so these rhythms can be expected to show a flexibility not observed with more primary rhythmic phenomena such as the cardiac cycle. Since temporal coordinated activities have their own primary function, one can also expect that function to be initiated in response to the appropriate environmental stimuli. Thus, if one were to measure a biochemical or physiological parameter over the period of a circa-rhythm in a normal living condition, the pattern would reflect not just the circa-clock effect, but also the effects of any interval timers and external stimuli which impinged on the system. Just this point was made by Moore-Ede and colleagues in their delightful book, *The Clocks that Time Us*. This is represented graphically in Figure 5.20.

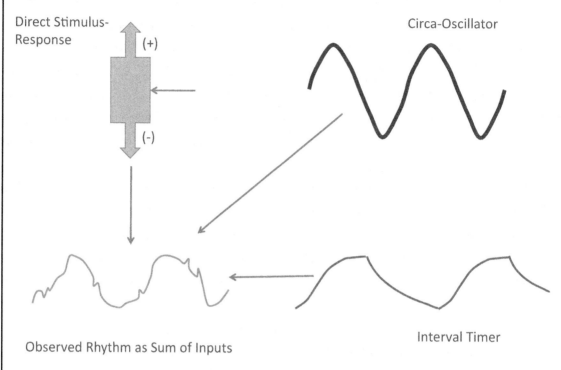

Figure 5.20: Influence of circa-oscillator, interval timer, and specific stimuli on observable rhythm. Note that the observed rhythm matches none of the inputs exactly but is a weighted sum of the various influences. The weighting will depend on the particular inputs, the function of the rhythmic factors and the evolutionary history of the organism. (Suggested by Moore-Ede, et al.[61].)

Of course, Figure 5.20 is based on a single, dominant circadian rhythm as the sole biological clock. We already know that this is not the case. Longer, seasonal cycles will interact with physiological and behavioral systems as well. In addition there are shorter cycles, ranging from many cycles per second (Hz) to minutes to hours, which also contribute to the final outcomes. Understanding the various contributions and interactions is a vital goal of research in biological rhythms.

5.7   OVERALL CONCLUSIONS

In Chapter 1, I argued that rhythms were a fundamental characteristic of living systems for three reasons. The first reason relates to the observation that many control systems in organisms rely on feedback control mechanisms that are inherently rhythmic. It takes very little in the way of stimulation to generate oscillations in such systems. The endocrine feedback system described in

Chapter 1 and action potentials in excitable tissue examined in Chapter 4 are only two examples of physiological systems demonstrating this cycling potential. Given the tight linkages between physiology and behavior on the one hand, and behavior and ecosystems on the other, only a few key rhythms are needed to create oscillations throughout the entire living world.

The second reason why rhythms are fundamental is the need to time, and thus coordinate, internal activities. In this, living systems are no different from any other goal-directed complex device. In order to achieve their objectives, complex devices must time internal events to:

a) Ensure resources are available to be allocated to that activity;

b) Prevent one activity from interfering with another.

Given that most activities are inherently repetitive (there are very few actions which occur only once), rhythms provide an efficient and effective means of maintaining the necessary internal temporal order.

The final reason ties the first two together into a coherent pattern. The existence of geophysical cycles—and the adaptive advantages of predicting associated temporal patterns—create a selection pressure for associating the internal temporal order with environmental cycles. In effect, the role of geophysical cycles is to reinforce internal temporal order through forced synchrony. In so doing, evolution has created both problems and opportunities with living systems that do not exist within most human-created devices.

The problem is as I described it in Chapter 1—organisms undergo a predictive series of physiological and behavioral changes throughout various geophysical timeframes, the most dominant being that of the day/night cycle. Organisms are, to put it simply, not the same at different times of day. Since the organism itself is not the same at different times, its interaction with its environment will not be the same either. After all, that is a primary reason for circadian rhythms—to alter the organism so that it maintains its level of adaptation to a changing external environment. Any approach to dealing with any biological organisms—including humans—that fails to adjust to this fundamental reality is going to be less effective than a time-varying approach. This reality applies to any productive human activity, whether in business, health care, education, engineering, or any other. It does not matter what the activity entails or what its purpose may be. If the same approach to conducting the activity is used at different times of day, the benefits will be less and the risk higher than if the proper temporal adjustments are made. No matter how sophisticated our lives and our jobs become, we still must confront basic evolutionary realities and physical constraints.

Rhythms do not merely create problems with which we must contend—they represent opportunities as well. Recall the discussion from Chapter 1 about the roles of the LD cycle and suprachaismatic nuclei (SCN) in maintaining rhythms in elderly human beings. That discussion may not have made much sense at the time, but let's reconsider it in the light of the materials presented in subsequent chapters. Consider the idea that the SCN represent master pacemakers for the human circadian system analogous to the role played by the SA and AV nodes (pacemakers) for the heart. The SCN are autorhythmic, help coordinate the body's other rhythms, and serve as an interface with the outside world. When cardiac pacemakers fail, we replace them with artificial pacemakers to maintain coherent rhythmicity. What if a significant aspect of aging is the progressive failure of our circadian pacemakers? Would not such a failure place the physiology and behavior of elderly humans at risk through internal desynchronization? And if so, would it not be prudent to correct this situation through the application of an artificial circadian pacemaker analogous to a cardiac pacemaker?

In fact, there might be a fairly simple approach to this problem. High amplitude light/dark cycles can help reinforce SCN rhythms. If we apply this principle to the elderly, then we should be exposing these individuals to high-amplitude LD cycles to help maintain rhythmic coherence, thus countering age-related weakening in the SCN. Theoretically, then, the proper application of lighting could retard those age-related changes exacerbated by internal desynchronization. In later chapters, we will examine other opportunities associated with biological rhythms.

The next chapter is a bit of a change-of-pace. Up to now, the mathematical side of oscillations has largely been ignored with the exception of using mathematical models to examine the plausibility of certain assertions and hypotheses. There is, however, a rich history in engineering involving oscillations, their characteristics, control, and interactions. With the inestimable assistance of Dr. Bahrad Sokhansanj, we will now turn our attention to that source of knowledge.

REFERENCES

[1] Aanonsen, A. (1959). Medical problems of shift work. *Industrial Medicine and Surgery* 28: 422–427. 145

[2] Akerstadt, T. and Gillberg, M. (1981). Sleep disturbance and shift work. In Reinberg, A., Vieux, N. and Andlauer, P. (eds.) *Advances in the Biosciences, Vol. 30: Night and Shift Work Biological and Social Aspects*. Pergamon Press: Oxford, Eng., pp. 127–137. 145

[3] Akerstadt, T. and Torsvall, L. (1980). Age, sleep, and adjustment to shift work. In Koella, W.P. (ed.) *Sleep 1980*. S. Karger: Basel, pp. 190–195. 145

[4] Albers, H., Gerall, A.A. and Axelson, J.F. (1981). Circadian rhythm dissociation in the rat: Effects of long-term constant illumination. *Neuroscience Letters* 25: 89–94. DOI: 10.1016/0304-3940(81)90106-3 138

[5] Athanassenas, G. and Walters, C.L. (1981). Sleep after transmeridian flights. In Reinberg, A., Vieux, N and Andlauer, P. (eds.) *Advances in the Biosciences, Vol. 30: Night and Shift Work: Biological and Social Aspects*. Pergamon Press: Oxford, Eng., pp. 139–147. 142

[6] Aschoff, J. (1969). Desynchronization and resynchronization of human circadian rhythms. *Aerospace Medicine* 40: 844–849. 141

[7] Aschoff, J. (1981). Freerunning and entrained circadian rhythms. In Aschoff, J. (ed.) *Handbook of Behavioral Neurobiology 4: Biological Rhythms*. Plenum Press: New York, pp. 81–93. 134, 138

[8] Ashoff, J. and Wever, R.A. (1976). Human circadian rhythms: A multioscillator system. *Federation of American Societies of Experimental Biology* 35: 2326–2332. DOI: 10.1038/nrg1633 141

[9] Baldessarini, R.J., Campbell, A., Madsen, J., Hershel, M., Finkelstein, S., Smith, J.M., Majocha, R. and Smith, J.M. (1981). Chronopharmacology. *Psychopharmacology Bulletin* 17(3): 112–113. 141

[10] Beljan, O., Winget, C. and Rosenblatt, L. (1973). The desynchronosis syndrome. In Annual Scientific Meeting: Aerospace Medical Association, Washington, D.C., Pre-prints of the scientific program, pp. 223. 142

[11] Boivin, D., Tremblay, G. and James, F. (2007). Working on atypical schedules. *Sleep Medicine* 8: 578–589. DOI: 10.1016/j.sleep.2007.03.015 145

[12] Carruthers, M., Arguelles, A.E. and Mosovich, A. (1976). Man in transit: Biochemical and physiological changes during intercontinental flights. *Lancet* 1(7967): 977–980. DOI: 10.1016/S0140-6736(76)91857-2

[13] Caruso, C.C., Lusk, S.L. and Gillespie, B.W. (2004). Relationship of work schedules to gastrointestinal diagnoses, symptoms, and medication use in auto factory workers. *American Journal of Industrial Medicine* 46(6): 586–598. DOI: 10.1002/ajim.20099 145

[14] Costa, G., Apostali, P. d'Andrea, F. and Gaffuri, E. (1981). Gastrointestinal and neurotic disorders in textile shift workers. In Reinberg, A., Vieux, N. and Andlauer, P. (eds.) *Advances in the Biosciences, Vol. 30: Night and Shift Work Biological and Social Aspects.* Pergamon Press: Oxford, Eng., pp. 215–221. 145

[15] Costa, G. (1996). The impact of shift and night work on health. *Applied Ergonomics* 27(1): 9–16. DOI: 10.1016/0003-6870(95)00047-X 145, 146

[16] Craig, L.A. and McDonald, R.J. (2008). Chronic disruption of circadian rhythms impairs hippocampal memory in the rat. *Brain Research Bulletin* 76(1–2): 141–151. DOI: 10.1016/j.brainresbull.2008.02.013 148, 149

[17] Davis, S. and Mirick, D.K. (2006). Circadian disruption, shift work and the risk of cancer: a summary of the evidence and studies in Seattle. *Cancer Causes and Control* 17: 539–545. DOI: 10.1007/s10552-005-9010-9 145

[18] Davis, S., Mirick, D.K. and Stevens, R.G. (2001). Night shift work, light at night and risk of breast cancer. *Journal of the National Cancer Institute* 93(20): 1557–1562. DOI: 10.1093/jnci/93.20.1557 145

[19] DeCoursey, P.J., Krulas, J.R., Mele, G. and Holley, D.C. (1997). Circadian performance of suprachiasmatic nuclei (SCN)-lesioned antelope ground squirrels in a desert enclosure. *Physiology and Behavior* 62: 1099–1108. DOI: 10.1016/S0031-9384(97)00263-1 149

[20] DeCoursey, P.J., Walker, J.K. and Smith, S.A. (2000). A circadian pacemaker in free-living chipmunks essential for survival? *Journal of Comparative Physiology A* 186: 169–180. DOI: 10.1007/s003590050017 149

[21] Endo, S., Yamamoto, T. and Sashi, M. (1981). Effects of time zone changes on sleep. In Johnson, C., Tepas, D.I., Colquhoun, W.P. and Colligan, M.J. (eds.) *Biological Rhythms, Sleep and Shift Work*. Spectrum Publications: Jamaica, New York, pp. 415–434. 142

[22] Enright, J.T. (1981). Methodology. In Aschoff, J. (ed.) *Handbook of Behavioral Neurobiology 4: Biological Rhythms*. Plenum Press: New York, pp. 11–19. 132

[23] Eriksson, M.E. and Millar, A.J. (2003). The circadian clock: A plant's best friend in a spinning world. *Plant Physiology* 132: 732–738. DOI: 10.1104/pp.103.022343

[24] Fan, J.-Y., Muskus, M.J., Preuss, F. and Price, J.L. (2008). Evolutionarily conserved features of vertebrate chi delta and *Drosophia* Dbt in the circadian mechanism. Presented at the Society for Research on Biological Rhythms 20th Anniversary Meeting, Destin, FL, May 17–21. Abstract 37, Pg 80.

[25] Fido, A. and Ghali, A. (2008). Detrimental effects of variable work shifts on quality of sleep, general health and work performance. *Medical Principles and Practice* 17(6): 453–457. DOI: 10.1159/000151566 145

[26] Filipski, E., Delaunay, F., King, V.M., Wu, M.-W., Claustrat, B., Grechez-Cassiau, A., Guettier, C., Hastings, M.H. and Levi, F. (2004). Effects of chronic jet lag on tumor progression in mice. *Cancer Research* 64: 7879–7885. DOI: 10.1158/0008-5472.CAN-04-0674 147

[27] Filipski, E., Innominato, P.F., Wu, MW., Li, X.-M., Iacobelli, S., Xian, L.-J. and Levi, F. (2005). *Journal of the National Cancer Institute* 97(7): 507–517. 147, 148

[28] Filipski, E., King, V.M., Li, XM., Mormont, M.-C., Claustrat, B., Hastings, M. and Levi, F. (2003). Disruption of circadian coordination accelerates malignant growth in mice. *Pathologie Biologie* 51(4): 216–219. DOI: 10.1016/S0369-8114(03)00034-8 147, 148

[29] Filipski, E., King, V.M., Li, XM., Granda, T.G., Mormont, M.-C., Liu, XH., Claustrat, B., Hastings, M.H. and Levi, F. (2002). Host circadian clock as a control point in tumor progression. *Journal of the National Cancer Institute* 94(9): 690–697. DOI: 10.1093/jnci/94.9.690 148

[30] Filipski, E., Li, X.M. and Levi, F. (2006). Disruption of circadian organization and malignant growth. *Cancer Causes and Control* 17: 509–514. DOI: 10.1007/s10552-005-9007-4

[31] Foret, J. and Benoit, O. (1976). Shift work and sleep. In Koella, W.P. and Levin, P. (eds.) *Sleep 1976*. Third European Congress of Sleep Research, Monpellier, 1976. S. Karger: Basel, pp. 81–86. 145

[32] Forger, D.B. and Paydarfar, D. (2004). Starting, stopping and resetting biological oscillators: in search of optimum perturbations. *Journal of Theoretical Biology* 230: 521–532. DOI: 10.1016/j.jtbi.2004.04.043 137

[33] Fujita, T., Mori, H., Minowa, M., Kimura, H., Tsujishita, J., Kimura, K., Goa, J., Yoshida, A., Morita, T., Mitsubayashi, M., et al. (1993). A retrospective cohort study on long-term effects of shift work. *Nippon Koshu Eisei Zasshi* 40(4): 273–283. 145

[34] Fujiwara, S., Shinkai, S., Kurokawa, Y. and Wataanabe, T. (1992). The acute effects of experimental short-term evening and night shifts on human circadian rhythm: the oral temperature, heart rate, serum cortisol and urinary catecholamine levels. *International Archives of Occupational and Environmental Health* 63(6): 409–418. DOI: 10.1007/BF00386937 141

[35] Ghiasvand, M., Heshmat, R., Golpira, R., Haghpanah, V., Soleimani, A., Shoushtarizadeh, P., Tavangar, S.M. and Larijani, B. (2006). Shift work and the risk of lipid disorders: A cross-sectional study. Lipids in Health and Disease, 5:9. (http://www.lipidworld.com/conent/5/1/9). DOI: 10.1186/1476-511X-5-9 145

[36] Hak, A. and Kampman, R. (1981). Working irregular hours: Complaints and state of fitness of railway personnel. In Reinberg, A., Vieux, N. and Andlauer, P. (eds.) *Advances in the Biosciences, Vol. 30: Night and Shift Work Biological and Social Aspects*. Pergamon Press: Oxford, Eng., pp. 229–236.

[37] Hansen, J. (2006). Risk of breast cancer after night- and shift work: current evidence and ongoing studies in Denmark. *Cancer Causes and Control* 17: 531–537. DOI: 10.1007/s10552-005-9006-5 145

[38] Harrington, J. (1994). Shift work and health—A critical review of the literature on working hours. *Annals of the Academy of Medicine, Singapore* 23(5): 699–705. 145

[39] Haus, E. and Smolensky, M. (2006). Biological clocks and shift work: Circadian dysregulation and potential long-term effects. *Cancer Causes Control* 17: 489–500. DOI: 10.1007/s10552-005-9015-4 145

[40] Hautz, G.T. and Adams, T. (1966). Phase shifts of the human circadian rhythm and performance deficit during periods of transition: I. East-west flights. *Aerospace Medicine* 37: 688–674. 142

[41] Healy, D., Minors, D.S. and Waterhouse, J.M. (1993). Shiftwork, helplessness and depression. *Journal of Affective Disorders* 29: 17–25. DOI: 10.1016/0165-0327(93)90114-Y 145

[42] Honma, K.-I. and Hiroshige, T. (1978). Internal desynchronization among several circadian rhythms in the rat under constant light. *American Journal of Physiology* 235(5): R243–R249. 138

[43] Iglesias, R., Terres, A. and Chavarria, A. (1980). Disorders of the menstrual cycle in airline stewardesses. *Aviation, Space and Environmental Medicine* 51(5): 518–520. 143

[44] Johnson, C.H. (2001). Endogenous timekeeping in photosynthetic organisms. *Annual Review of Physiology* 63: 695–728. DOI: 10.1146/annurev.physiol.63.1.695

[45] Katz, G., Durst, R., Zislin, Y., Barel, Y. and Knobler, H.Y. (2001). Psychiatric aspects of jet lag: Review and hypothesis. *Medical Hypotheses* 56(1): 20–23. DOI: 10.1054/mehy.2000.1094 142

[46] Katz, G., Knobler, H.Y., Laibel, Z., Strauss, Z. and Durst, R. (2002). Time zone change and major psychiatric morbidity: The results of a 6-year study in Jerusalem. *Comprehensive Psychiatry* 43(1): 37–40. DOI: 10.1053/comp.2002.29849 142

[47] Klein, K.E., Bruner, H., Holtzman, H., Relme, H., Stolze, J., Steinhoff, W.D. and Wegmann, H.M. (1970). Circadian rhythms of pilots' efficiency and effects of multiple time zone travel. *Aerospace Medicine* 41: 125–132. 142

[48] Klein, K.E., Bruner, H., Gunther, E., Jovy, D., Mertens, J., Rinpler, A. and Wegmann, H.M. (1972). Psychological and physiological changes caused in desynchronization following transzonal air travel. In Colquhoun, W.P. (ed.) *Aspects of Human Efficiency*. English Universities Press., Ltd: London, pp. 295–305. 141, 142

[49] Klein, K.E., Wegmann, H.M. and Hunt, B.I, (1972). Desynchronization of body temperature and performance circadian rhythms as a result of outgoing and homegoing transmeridian flights. *Aerospace Medicine* 43(2): 119–132. 142

[50] Klein, K.E. and Wegmann, H.M. (1974). The resynchronization of human circadian rhythms after transmeridian flight as a result of flight direction and mode of activity. In Scheving, L., Halberg, F. and Pauley, J. (eds.) *Chronobiology*. Elgaku Shoin, Ltd: Tokyo, pp. 564–570. 142

[51] Koller, M. (1983). Health risks related to shift work. An example of time-contingent effects of long-term stress. *International Archives of Occupational and Environmental Health* 53(1): 59–75. DOI: 10.1007/BF00406178

[52] Koller, M., Haider, M. Kundi, M. Cervinka, R. Katschig, H. and Kufferle, B. (1981). Possible relations of irregular working hours to psychiatric psychosomatic disorders. In Reinberg, A., Vieux, N. and Andlauer, P. (eds.) *Advances in the Biosciences, Vol. 30: Night and Shift Work Biological and Social Aspects*. Pergamon Press: Oxford, Eng., pp. 465–472. 145

[53] Knuttson, A. (2003). Health disorders of shift workers. *Occupational Medicine* 53: 103–108. DOI: 10.1093/occmed/kqg048 145

[54] Knutsson, A., Hallquist, J., Reuterwall, C., Theorell, T. and Akerstadt, T. (1999). Shiftwork and myocardial infarction: a case-control study. *Occupational and Environmental Medicine* 56: 46–50. DOI: 10.1136/oem.56.1.46 145

[55] Kripke, D.F., Cook, B. and Lewis, O.F. (1971). Sleep of night workers: EEG recordings. *Psychophysiology* 7(3): 377–384. DOI: 10.1111/j.1469-8986.1970.tb01762.x

[56] Larsen, K.R., Barattini, P., Dayton, M.T. and Moore, J.G. (1994). Effects of constant light on rhythmic gastric function in fasting rats. *Digestive Diseases and Sciences* 39(4): 678–688. DOI: 10.1007/BF02087408

[57] Martino, T.A., Tata, N., Belsham, D.D., Chalmers, J., Straume, M., Lee, P., Pribiag, H., Khaper, N., Liu, P.P., Dawood, F., Backx, P.H., Ralph, M.R. and Sole, M.J. (2007). Disturbed diurnal rhythm alters gene expression and exacerbates cardiovascular disease with rescue by resynchronization. *Hypertension* 49: 1104–1113. DOI: 10.1161/HYPERTENSIONAHA.106.083568 147

[58] Martino, T.A., Oudit, G.Y., Herzenberg, A.M., Tata, N., Koletar, M.M., Kabir, G.M., Belsham, D.D., Backx, P.H., Ralph, M.R. and Sole, M.J. (2008). Circadian rhythm disorganization produces profound cardiovascular and renal disease in hamsters. *American Journal of Physiology-Regulatory, Integrative and Comparative Physiology* 294: 1675–1683 DOI: 10.1152/ajpregu.00829.2007

[59] Moore, J.G., Larsen, K.R., Barattini, P., and Dayton, M.T. (1994). Asynchrony in circadian rhythms of gastric function in the rat: A model for gastric mucosal injury. *Digestive Diseases and Sciences* 39(8): 1619–1624. DOI: 10.1007/BF02087766

[60] Moore-Ede, M. and Richardson, G. (1985). Medical implications of shift work. *Annual Review of Medicine* 36: 607–617. DOI: 10.1146/annurev.me.36.020185.003135 145

[61] Moore-Ede, M.C., Sulzman, F.M. and Fuller, C.A. (1982). *The Clocks that Time Us*. Harvard University Press: Cambridge, MA. 141, 160

[62] Paranjpe, D.A. and Sharma, V.K. (2005). Evolution of temporal order in living organisms. *Journal of Circadian Rhythms* 3:7. Doi:10.1186/1740--3391-3-7. (http://www.jcircadianrhythms.com/content/3/1/7). DOI: 10.1186/1740-3391-3-7 133

[63] Pavlidis, T. (1981). Mathematical models. In Aschoff, J. (ed.) *Handbook of Behavioral Neurobiology 4: Biological Rhythms*. Plenum Press: New York, pp. 41–54. 137

[64] Penev, P., Kolker, D., Ze, P. and Turek, F. (1998). Chronic circadian desynchronization decreases the survival of animals with cardiomyopathic heart disease. *American Journal of Physiology* 275(6 pt. 2): H2334–H2337. 147

[65] Pietroiusti, A., Forlini, A., Magrini, A., Galante, A., Coppeta, L., Gemma, G., Romeo, E. and Bergamaschi, A. (2008). Shift work increase the frequency of duodenal ulcer in *H. pylori* infected workers. *Occupational and Environmental Medicine* 63: 773–775. DOI: 10.1136/oem.2006.027367 145

[66] Pikovshy, A., Rosenblum, M. and Kurths, J. (2003). *Synchronization: A Universal Concept in Nonlinear Sciences*. Cambridge University Press: Cambridge, UK. 137

[67] Pittendrigh, C.S. (1981a). Circadian rhythms: General perspective. In Aschoff, J. (ed.) *Handbook of Behavioral Neurobiology 4: Biological Rhythms*. Plenum Press: New York, pp. 57–80. 141

[68] Pittendrigh, C.S. (1981b). Circadian rhythms: Entrainment. In Aschoff, J. (ed.) *Handbook of Behavioral Neurobiology 4: Biological Rhythms*. Plenum Press: New York, pp. 95–124. 134, 141

[69] Plamen, D., Kolker, D.E., Zee, P.C. and Turek, F. (1998). Chronic circadian desynchronization decreases the survival of animals with cardiomyopathic heart disease. *American Journal of Physiology—Heart and Circulatory Physiology* 275: 2334–2337.

[70] Preston, F.S., Bateman, C.S., Short, R.V. and Wilkinson, R.T. (1973). Effects of flying and time changes on menstrual cycle length and on performance in airline stewardesses. *Aerospace Medicine* 44: 438–4443. 143

[71] Preuss, F., Tang, Y., Laposky, A.D., Arble, D., Keshavarzian, A. and Turek, F. (2008). Adverse effects of chronic circadian desynchronization in animals in a "challenging" environment. *American Journal of Physiology-Regulatory, Integrative and Comparative Physiology* (Epub ahead of print). DOI: 10.1152/ajpregu.00118.2008 147

[72] Rea, M.S., Bierman, A. Figueiro, M.G. and Bullough, J.ED. (2008). A new approach to understanding the impact of circadian disruption on human health. *Journal of Circadian Rhythms* 6:7. http://www.jciracdianrhythms.com/content/6/1/7. DOI: 10.1186/1740-3391-6-7 141

[73] Refinetti, R. (2006). *Circadian Physiology, 2nd Edition*. CRC Taylor and Francis: Baca Raton, FL, pp. 529–539. 127

[74] Reinberg, A, Andluaer, P. and Vieux, N. (1981). Circadian temperature rhythm amplitude and long term tolerance to shiftworking. In Johnson, C.,, Tepas, D.I., Colquhoun, W.P. and Colligan, M.J. (eds.) *Biological Rhythms, Sleep and Shift Work*. Spectrum Publications: Jamaica, New York, pp. 61–74.

[75] Roenneberg, T. and Merrow, M. (2005). Circadian clocks: The fall and rise of physiology. *Nature Reviews* 6: 965–971. DOI: 10.1038/nrm1766 133

[76] Rogers, H.L. and Reilly, S.M. (2002). A survey of health experiences of international business travelers: Part one—Physiological aspects. *American Association of Occupational Health Nurses Journal* 50(10): 449–459. 142

[77] Rosbash, M. (1995). Molecular control of circadian rhythms. *Current Opinion in Genetics and Development* 5: 662–668. DOI: 10.1016/0959-437X(95)80037-9

[78] Rudiger, H.W. (2004). Health problems due to night shift and jetlag. *Internist* 45(9): 1021–1025. DOI: 10.1007/s00108-004-1257-9 143

[79] Sasaki, M. and Endo, S. (1985). Disturbances of the circadian sleep-wake rhythm after time zone changes. *Journal of UOEH* 7: Suppl: 141–150. 142

[80] Schernhammer, E.S., Laden, F., Speizer, F.E., Willett, W.C., Hunter, D.J., Kawachi, I.K., Fuchs, C.S. and Colditz, G.A. (2003). Night-shift work and risk of colorectal cancer in the nurses' health study. *Journal of the National Cancer Institute* 95(11): 825–828. DOI: 10.1093/jnci/95.11.825 145

[81] Schernhammer, E.S., Laden, F., Speizer, F.E., Willett, W.C., Hunter, D.J., Kawachi, I.K. and Colditz, G.A. (2001). Rotating night shifts and risk of breast cancer in women participating in the nurses' health study. *Journal of the National Cancer Institute* 93(20): 1563–1568. DOI: 10.1093/jnci/93.20.1563 145

[82] Spitzer, R.L., Terman, M., Williams, J.B.W., Terman, J.S., Malt, U.F., Singer, F. and Lewy, A.J. (1999). Jet lag: Clinical features, validation of a new syndrome-specific scale, and lack of response to melatonin in a randomized, double-blind trial. *American Journal of Psychiatry* 156: 1392–1396. 142

[83] Su, S.B., Lu, C.W., Kao, Y.Y. and Guo, H.R. (2008). Effects o 12-hour rotating shifts on menstrual cycles of photoelectronic workers in Taiwan. *Chronobiology International* 25(2): 237–248. DOI: 10.1080/07420520802106884 145

[84] Suwazono, Y., Dochi, M., Sakata, K., Okubo, Y., Oishi, M., Tanaka, K., Kobayashi, E. and Nogawa, K. (2008). Shiftwork is a risk factor for increased blood pressure in Japanese men. A 14-year historical cohort study. *Hypertension* 52: 581–586. DOI: 10.1161/HYPERTENSIONAHA.108.114553 145

[85] Sveinsdottir, H. (2006). Self-measured quality of sleep, occupational health, working environment, illness experience and job satisfaction of female nurses working different combination of shifts. *Scandinavian Journal of Caring Science* 20: 229–237. DOI: 10.1111/j.1471-6712.2006.00402.x 145, 146

[86] Taub, J.M. and Berger, R.J. (1973). Sleep stage patterns associated with acute shifts in the sleep-wakefulness cycle. *Electroencephalography and Clinical Neurophysiology* 35: 613–619. DOI: 10.1016/0013-4694(73)90214-9 143

[87] Taub, J.M. and Berger, R.J. (1974). Acute shifts in the sleep-wakefulness cycle: Effects on performance and mood. *Psychosomatic Medicine* 36 (2): 164–173. 143

[88] Taylor, P.J., Pocock, S.J. and Sergean, R. (1972a). Absenteeism of shift and day workers. *British Journal of Industrial Medicine* 29: 208–213. 146

[89] Taylor, P.J. and Pocock, S.J. (1972b). Mortality of shift and day workers 1956–1968. *British Journal of Industrial Medicine* 29: 201–207. 146

[90] Tec, L. (1981). Depression and jet lag. *American Journal of Psychiatry* 138:858. 142

[91] Tobia, L., Spera, G., Cruciana, D., Fanelli, C., Diana, S., Spagnoli, F., Necozione, S. and Paoletti, A. (2007). Occupational risk stress in shift workers. *Giornale Italiano di Medicina del Lavoro ed Ergomania* 29(3 Siuppl.): 700–701.

[92] Tuchsen, F, Hannerz, H. and Burr, H. (2008). A 12 year prospective study of circulatory disease among Danish shift workers. *Occupational and Environmental Medicine* 63: 451–455. DOI: 10.1136/oem.2006.026716 145

[93] van der Heiligenberg, S., Depres-Brummer, P., Barbason, H., Claustrat, B., Reyenes, M. and Levi, F. (1999). The tumor promoted effect of constant light exposure on diethylnitrosamine-induced hepatocarcinogenesis in rats. *Life Sciences* 64(26): 2523–2534. DOI: 10.1016/S0024-3205(99)00210-6 147

[94] Verhaegen, F., Maasen, A. and Meers, A. (1981). Health problems in shift workers. In Johnson, C.,, Tepas, D.I., Colquhoun, W.P. and Colligan, M.J. (eds.) *Biological Rhythms, Sleep and Shift Work*. Spectrum Publications: Jamaica, New York, pp. 271–282. 145

[95] Wager-Smith, K. and Kay, S.A. (2000). Circadian rhythm genetics: from flies to mice to humans. *Nature Genetics* 26: 23–27. DOI: 10.1038/79134

[96] Walsh, J.K., Tepas, D.I. and Moss, P.D. (1981). The EEG of night and rotating shift workers. In Johnson, C.,, Tepas, D.I., Colquhoun, W.P. and Colligan, M.J. (eds.) *Biological Rhythms, Sleep and Shift Work*. Spectrum Publications: Jamaica, New York, pp. 371–381. 145

[97] Weitzman, E.D. and Kripke, D.F. (1981). Experimental 12-hour shift of the sleep-wake cycle in man: Effects on sleep and physiologic rhythms. In Johnson, C., Tepas, D.I., Colquhoun, W.P. and Colligan, M.J. (eds.) *Biological Rhythms, Sleep and Shift Work*. Spectrum Publications: Jamaica, New York, pp. 93–110. 145

[98] Wever, R.A. (1979). *The Circadian System of Man: Results of Experiments under Temporal Isolation*. Springer-Verlag: New York. 141

[99] Wever, R.A. (1983). Fractional desynchronization of human circadian rhythms: A method for evaluating entrainment limits and functional interdependencies. *Pflugers Archives* 396: 128–137. DOI: 10.1007/BF00615517 142

[100] Winfree, A.T. (1987). *The Timing of Biological Clocks*. Scientific American Books: New York. 137

[101] Yadegarfar, G. and McNamee, R. (2008). Shift work, confounding and death from ischaemic heart disease. *Occupational and Environmental Medicine* 65: 158–163. DOI: 10.1136/oem.2006.030627 145

CHAPTER 6

The Circle Game: Mathematics, Models, and Rhythms

Overview

In this chapter, Dr. Bahrad Sokhansanj comes to my aid to help introduce the fundamentals *of mathematical modeling* as applied to *biological oscillations*, using circadian rhythms as an example. The basics of mathematical modeling as applied to biological systems is briefly discussed, followed by an introduction to *linear models* applied to oscillating systems. The models are initially applied to mechanical devices to show how different values can lead to either damped or sustained oscillations. The concept of a *phase plane* or space as a graph of potential oscillatory behavior is then introduced. Next, the limits of linear models are explored, leading to the introduction of *non-linear models* beginning with *Van der Pol* models and introducing the concept of a *limit cycle*. These models are then *applied to biological systems*, especially subcelluar systems such as *transcription-feedback loops*, to ascertain the extent to which mathematical models approach living reality. *External effects*, such as selected environmental cycles and perturbations, are examined as factors impacting the models. As a result, new insights are gained concerning the *nature of entrainment* and the effects of the external environment on biological rhythms.

Basically, the point of making models is to be able to bring a measure of order to our experiments and observations, as well as to make scientific predictions about certain aspects of the world we experience.

-John Casti, Reality Rules (1992)

The sciences do not try to explain, they hardly even try to interpret, they mainly make models. By a model is meant a mathematical construct which, with the addition of certain verbal interpretations describes observed phenomena. The justification of such a mathematical construct is solely and precisely that it is expected to work.

- John von Neumann

We must use time as a tool not as a couch.

- John Fitzgerald Kennedy

And the seasons, they go round and round;
And the painted ponies go up and down;

We're captive on a carousel of time;

We can't return, we can only look;

Behind from where we came;

And go round and round and round;

In the circle game.

– Joni Mitchell, The Circle Game (1970)

6.1   INTRODUCTION TO MATHEMATICAL MODELING

The preceding chapters should have convinced you that biological rhythms are not the product of unique, singular processes at the molecular, cellular, or physiological level but rather arise due to the function of systems, combinations of processes that work together at multiple levels of organization. In this chapter, we will discuss some of the basic principles that allow quantitative scientists and engineers to describe and predict the dynamics of systems using mathematical equations. We will not be going into all the details of the theory behind those dynamics or the ways in which they are used in applications. Instead, we will try to illustrate key principles that are used in the contemporary modeling literature for systems that generate biological rhythms.

Most of these examples from the literature focus on circadian rhythms and the cellular processes that generate them. As a result, we also focus on circadian cycles. This should not be taken to mean that circadian rhythms are the only rhythms that matter or even the only circarhythms of importance. However, circadian rhythms do display certain characteristics which, if properly understood, can be applied more generally. These characteristics include autorhythmicity and the ability to be synchronized to another cycle. In the case of circadian rhythms, the latter includes the process of entrainment to the geophysical cycle of day and night. By examining circadian rhythms, we can learn much about the general principles underlying both autorhythmicity and entrainment.

In general, biological systems are made up of atoms, molecules, and chemical reactions. The difference between cells growing in a Petri dish and chemicals mixed in a test tube is one of *complexity*. There are millions of different kinds of molecules in cells and tissues. The activities of those cells and tissues are performed and regulated through millions of chemical reactions. Those reactions are connected together in multiple, overlapping *reaction networks*. This complexity poses significant challenges to the use of theoretical concepts from physics and chemistry to describe biological processes.

Instead of formal reaction models and equations to describe a biological system, biologists more typically use verbal and diagrammatic descriptions. Why do biologists use words and pictures instead of numbers and equations? Biological systems certainly follow the laws of physics and chemistry. So, we could in principle use mathematical models to analyze biological systems, much as we do with complex mechanical and chemical systems like engines, atmospheric pollution, or bridge structures. However, unlike physical phenomena in a controlled particle accelerator and chemical reactions in a test tube, the activity of biological systems is difficult to measure with quantitative accuracy.

Biologists resort to *qualitative* (word and picture) descriptions, because the experimental techniques they use are inevitability indirect and noisy.

Although it may not seem like it, the systems capable of mathematical descriptions in physics and chemistry are actually fairly simple. Modelers can therefore analyze the organization or structure with limited numbers of variables. However, biological systems are never simple and inevitably require more complex mathematical formulations. Indeed, the complexity of biological systems results in so-called *emergent* properties—characteristics that are the result of complex interactions not predictable simply from the behavior of the system components when considered in isolation. Our topic area, rhythmicity, is one of these emergent properties.

Nevertheless, we can work toward developing mathematical models provided we make certain assumptions. Based upon these assumptions, we can reduce the size of systems analyzed to the interaction of a limited number of variables that can be reliably measured. By focusing on a few key genes and hormones, modelers develop simplified representations of the system that can be analyzed mathematically to derive key behavioral principles. For example, we can ask how chemical reactions catalyzed by proteins produced in individual cells can give rise to synchronized oscillations throughout a cluster of cells in the human heart or brain.

Why is it so important to analyze systems using mathematical models? Mathematical models are formal ways of describing systems through equations and other rigorous relationships (such as logical rules) that describe the behavior of variables that represent system properties. It is important to note that not all system models are rigorous and use mathematical equations written down on paper or entered in a computer. The words and pictures shown throughout most of this text actually represent "models" of biological rhythm-generating systems as well. However, such conceptual models require some form of verification to estimate how accurate and/or useful they are in estimating real-world phenomena. In fact, such mathematical formulations can generate new hypotheses about system function, as we saw in previous chapters. That is why we converted the endocrine feedback and neuron models into mathematical constructs.

A model can even be physically constructed. A famous example of that is the modeling done by Watson and Crick to predict the structure of DNA molecules by determining what geometric arrangement of bonds could satisfy constraints identified through measurements. This was a kind of mathematical model, and today similar models are generated on computers and visualized on screens using 3-D rendering software. The models Watson and Crick constructed were effective because of their accuracy. Thus, if a particular geometric arrangement failed to predict measured properties, such as from data obtained by crystallography done by Rosalind Franklin, then that specific model could be rejected.

In general, though, while word and picture models are useful for describing concepts, they cannot be rigorously falsified through comparison with quantitative data. For example, there would in principle be no way of differentiating a *single oscillator* or *coupled multioscillator* model with just words. However, mathematical models of different oscillator configurations could be used to predict differences in their behavior that may be used to falsify one or the other case. In addition, they could be used to point out gaps in our knowledge and suggest new experiments.

In this chapter, we will give a brief overview of mathematical methods used to describe the dynamics of oscillating systems. This is intended as a review and primer for entering the field of biological rhythm modeling. What we will focus on here are the mathematical principles underlying these models and, in particular, the language of dynamical system modeling. Of course, as we learn more about the genes, proteins, and cell-cell interactions that are relevant for the generation and control of biological rhythms, the models used to describe these systems will grow ever more complex. Eventually, computer simulations will be necessary to visualize model predictions. However, the concepts we review here will help interpret the results of computer simulations and understand how their output can lead to biological implications.

6.2   LINEAR MODELS OF OSCILLATORS

Watson and Crick's wire and ball models of DNA and the diagrams shown in the other chapters of this textbook represent *abstractions* of the systems they describe. Certain aspects are selected as variables that can be represented explicitly as modifiable quantities in the model. Similarly, in a mathematical model, we select variables that are identified as being relevant for a system—certain proteins, for example, or important quantities like calcium ion concentration. The rules regarding the interaction of these variables are also abstracted in mathematical models. Not every physical interaction between variables need be, or even can be represented by an individual equation. Variables may also be related in an equation even if they do not directly interact, since they may interact through intermediaries that are not included in the system abstraction used to develop the model.

Parameters (constants) in the model may not correspond to a physically real quantity, such as the rate constant of a particular enzymatic reaction. Rather, they may represent multiple rates "lumped" together. For example, we may write down an equation that says the rate of growth of the number of bacteria (N) in a flask depends on the concentration of glucose (G) in that flask:

$$\frac{dN}{dt} = kG$$

Here, the rate of change of the number of bacteria (ΔN) per unit time (Δt) is represented by differential terms, since the change is assumed to be continuous (a common assumption that we will discuss in more detail toward the end of the chapter). In this equation, we assume that the dependence of rate of change on glucose level is *linearly* proportional (so G is not raised to any power—if we wanted to assume a quadratic proportionality, we would write $dN/dt = kG^2$, and so on). In this equation, the parameter relating the glucose concentration to bacterial growth rate is simply a single number k.

We could determine a numerical value for this parameter by performing linear regression—essentially, fitting a line to the data—on experimental data collected for bacterial growth rate versus glucose concentration. However, the actual process of bacteria consuming glucose and converting it into a changed growth rate (by producing proteins to drive cell replication) involves a large number of chemical reactions, all of which have particular rate constants. In this simple model, we have lumped the parameters together into a single value that can be estimated based on the available data. Because of the difficulty in measuring biological quantities, it is often necessary to do these kinds of simplifications to study complex cell systems.

As an example, consider the data represented in Figure 6.1a. The graph displays a linear relationship obtained through experimentation between the growth rate of yeast (*Zygosaccharomyces rouxii*) and glucose concentration in the growth medium, which can be represented by a simple equation such as described above. However, the underlying biochemistry is quite complex, as evidenced by the diagram of the glycolytic pathway provided in Figure 6.1b. Glycolysis alone represents but a small part of the hundreds of reactions needed to support cell division. So this mathematical model is clearly a highly abstract and simplified description of biological events.

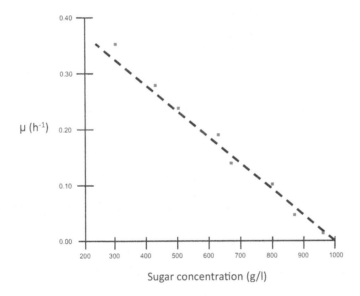

Sugar concentration (g/l)

Figure 6.1: Relationship between mathematical models and underlying biochemical processes: Growth rate versus glucose concentration in yeast. A. Model of growth rate in relationship to glucose concentration (dotted line). (Redrawn from Membre and colleagues[13].) *(Continues.)*

At the same time, such a model can still provide a measure of predictive validity by allowing for reasonable estimates of system behavior (in this case, growth). Thus, a mathematical formulation can convey useful information concerning the behavior of the biological system despite a lack of knowledge of the underlying details.

Typically, mathematical models of biological rhythms are based on the general principles of modeling dynamical systems. The concept underlying dynamical system models is that the physics

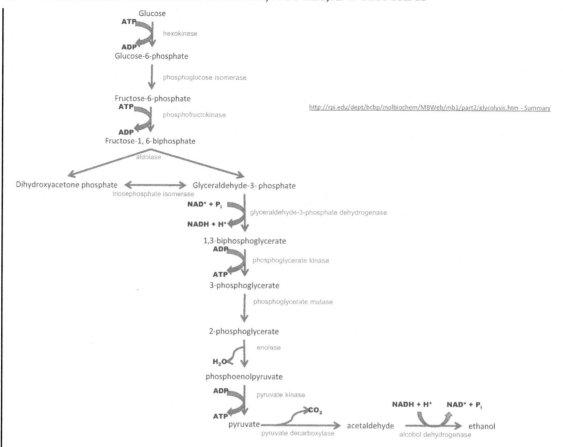

Figure 6.1: *(Continued.)* Relationship between mathematical models and underlying biochemical processes: Growth rate versus glucose concentration in yeast. B. Main reactions of glycolysis and fermentation.

behind the activity of objects can be rigorously described with mathematical equations. For all practical purposes, the movement of masses in response to forces can be accurately described by the relatively simple mathematical formulations of mechanical laws described by Gallileo, Newton, Kepler, and others in the 16th to 18th centuries. Of course, at certain limits these descriptions become incomplete—hence the 20th century development of quantum mechanics and relativity for describing systems at the scale of the very small and the very large. Even for systems they can describe accurately, there are limits to the applicability of dynamical system models, as we will see.

Dynamical system models were originally used to describe mechanical systems—masses, pulleys, pendulums—but they can be used to describe the dynamics of systems in general, including electrical circuits and, in the case of biological rhythms, molecular processes within cells. This means that the mathematical models we develop are based on an analogy between the *components* of a *biological* system and the *components* of a *mechanical* system.

To illustrate the fundamental concepts of dynamical system models, let's examine a highly abstracted system that can produce damped oscillations. We can model this system with a *linear* dynamical model that can be solved precisely using calculus. In general, linear systems that are more complex, as well as almost all *nonlinear* system models, cannot be solved so easily. However, the behavior of the systems can be predicted using computer simulations, and their qualitative behavior (e.g., ability to produce oscillations) can be described using approximations to the relatively basic mathematics we derive below.

Figure 6.2 shows a physical realization of the basic linear dynamical system model. A mass is suspended by a spring, which provides oscillations in the motion of the mass. A damper is included, which represents dissipative forces, such as friction in the spring. If we wanted to use this system to represent biological rhythms, we could develop an analogy where y, the displacement of the mass in Figure 6.2, represents the level of an oscillating biological substance, and the parameters k, c, and m of the system are lumped parameters that represent biochemical features that we do not include in detail at this level of abstraction. This is exactly analogous to lumping together the various parameters of biochemical reactions governing cell division in the example we discussed concerning microbial population growth.

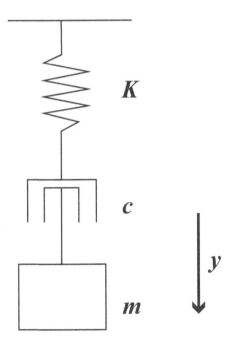

Figure 6.2: A mass-spring-damper system.

Using Newton's First Law, the force is equal to mass times acceleration (second derivative of the displacement y with respect to time t), which is countered by the restorative force of the spring (by Hooke's Law, proportional to its displacement) and the dissipative forces in the damper (i.e., friction, proportional to the velocity of the mass):

$$m\frac{d^2y}{dt^2} = ky - c\frac{dy}{dt} \tag{6.1}$$

Using the conventional notation $y' = dy/dt$ and $y'' = d^2y/dt^2$ and using some algebra to rearrange the equation, we obtain:

$$y'' + (c/m)y' + (k/m)y = 0 \tag{6.2}$$

We can write this as a linear system of ordinary differential equations (ODEs) if we let $y_1 = y$ and $y_2 = y'$:

$$\mathbf{y}' = \begin{bmatrix} y_1' \\ y_2' \end{bmatrix} = \begin{bmatrix} 0 & 1 \\ -k/m & -c/m \end{bmatrix} \begin{bmatrix} y_1 \\ y_2 \end{bmatrix} = \mathbf{A}\mathbf{y}' \tag{6.3}$$

where

$$\mathbf{A} = \begin{bmatrix} 0 & 1 \\ -k/m & -c/m \end{bmatrix}$$

Thus, this linear mathematical model can be expressed as a linear system of equations, which makes it easy to solve. You may recall from calculus classes that the solution of the scalar ODE for exponential growth $y' = ay$, where a is a constant growth rate, is the exponential function with respect to time, $y(t) = c\exp(at)$ where c is given by the initial amount of y. Similarly, the solution for a linear system of ODEs as described in (6.3) is given by a superposition of solutions with the form $\mathbf{y} = \mathbf{x}\exp(\lambda t)$, for a 2-D system,

$$\mathbf{y} = c_1\mathbf{x}^{(1)}\exp(\lambda_1 t) + c_2\mathbf{x}^{(2)}\exp(\lambda_2 t) \tag{6.4}$$

With some algebra, it can be shown that these solutions must satisfy the condition $\det(\mathbf{A} - \lambda\mathbf{I}) = 0$. That is, the terms in (6.4) correspond to the eigenvalues $\{\lambda_1, \lambda_2\}$ and eigenvectors $\{\mathbf{x}^{(1)}, \mathbf{x}^{(2)}\}$ of $\mathbf{A}$. The constants will be set by initial values of y_1 and y_2 at time $t = 0$. In the case of the linear mass-spring-damper system, the eigenvalues are the solutions to the quadratic equation:

$$\lambda^2 + (c/m)\lambda + k/m = 0$$
$$\lambda_{1,2} = -c/2m + (1/2)\sqrt{c^2/m^2 - 4k/m} \tag{6.5}$$

It turns out that the time-dependent evolution of the system will be controlled by the nature of the eigenvalues determined by Equation (6.5). Assuming that $c^2 < 4km$, the eigenvalues $\lambda_{1,2}$ will be complex, and if $c = 0$, the eigenvalues will be purely imaginary. In the latter case, the solutions $\mathbf{y}$ = [y_1 y_2] will be sinusoidal, representing indefinite oscillations. In physical terms, this represents the behavior of a frictionless spring: it will oscillate forever at constant amplitude. On the other hand, if the eigenvalues have a real part as well as an imaginary one, then the solutions will be a product of negative exponentials and sinusoidal functions. This corresponds to damped oscillations that decay in amplitude over time. This is what we would see physically for a spring with friction. It is also what we saw when we removed a periodic stimulus to our endocrine feedback model described in Chapter 1.

To display the behavior of a dynamical system model, it can be useful to plot solutions of the model equations as a "phase portrait," or graph in the "phase plane." While it is possible to plot a given solution for **y** versus time, the graph will look different for different initial values of **y** at time $t = 0$, which will lead to different values for the eigenvectors **x**. These differences will take the form of different quantitative values for amplitude of oscillations and rates at which the curve damps.

However, if we are interested in the *qualitative* behavior of the model, we are more concerned in the question of whether—for given sets of parameters—there are any oscillations at all, or if there are damped oscillations, or if the magnitude of **y** reaches a steady state at later times. The phase portrait of solutions **y** accomplishes this goal by plotting y_2 versus y_1 curves for a few different choices of initial conditions—enough to show the relevant qualitative information for system dynamics. Figure 6.3 shows examples of how in the $y_1 - y_2$ phase plane, the phase portrait of undamped oscillations are circular, and damped oscillations are spirals. These solutions for different initial conditions are called *trajectories* (sometimes called *orbits*) in the phase plane.

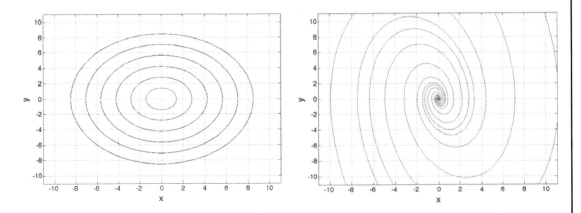

Figure 6.3: [Left] Circular trajectories in a phase plane, corresponding to $c = 0$ and $k/m = 1$ in Equation (6.3). [Right] Spiral trajectories representing oscillations in time with amplitude decaying to zero, corresponding to $c/m = 1$ and $k/m = 4$.

The node at $y_1 = y_2 = 0$ represents what is called either the *equilibrium* or *critical point* of the system (as with too many terms in mathematics, they are often used interchangeably; we will use the more descriptive "equilibrium point"). It is where $dy_1/dt = dy_2/dt = 0$. In the case of the spiral phase portrait of damped oscillatory solutions, it can be interpreted as a *steady state* of the system. As time increases toward infinity, the system eventually converges to the point where y_1 and y_2 (representing the displacement and speed of the mass in Figure 6.1 or an analogous abstraction) approach zero. In the case of the circular phase portrait of an oscillatory solution, it is the value around which the solution oscillates, though it never converges to that point. In both cases we call this a *stable* equilibrium or critical point, because the behavior of the system about this node is essentially static with respect to the initial conditions $\mathbf{y}(t = 0)$. In the case of spirals (damped oscillations), every solution ("trajectory") converges to zero, in the other they all oscillate about zero.

This situation arises mathematically due to the values of the eigenvalues $\lambda_{1,2}$ found by the Equation (6.5). Since they are physical quantities, the mass, spring, and damper parameters, m, k, and c, take on positive values. Thus, the eigenvalues calculated using Equation (6.5) are guaranteed to either be pure imaginary (giving rise to circular phase trajectories) or are complex with negative real components (giving rise to spirals that end up at the equilibrium point). If the eigenvalues were complex with positive real parts, the exponential terms would have positive coefficients, and there would be exponential growth in amplitude of oscillations rather than exponential decay. This is impossible for the mass-spring-damper system, because in physical reality there is energy conservation, and the system oscillations could not grow indefinitely. But equations that are used to describe other dynamical systems may involve approximations that do not explicitly include conservation principles. In that case, if the system eigenvalues determined by Equation (6.5) have positive real parts, then the equilibrium (or critical) point is *unstable*. Rather than trajectories spiraling in toward the node as shown in Figure 6.3, the trajectories would spiral outward and the amplitude of the solutions $\mathbf{y}$ would indeed grow indefinitely as time moves forward.

In examining Figure 6.3, it might be helpful to imagine how the data would look if you followed along a specific line. For example, following along one of the circles in the left-hand graph, the value for the amplitude would increase then decrease toward zero, increase in a negative direction, then rise toward zero, then increase and so on forever. This creates a *sustained oscillation*. On the other hand, doing the same thing using any line in the right-hand graph, the amplitude rises and falls but always to a lesser amount, eventually reaching zero. This is a *damped oscillation*. The physical explanation is that sustained oscillations are only possible in this model in that there are no dissipative forces, i.e., $c = 0$.

Since we have strayed pretty far from biology, it is important to remember that while there is no physical "spring" or "damper" in circadian rhythm systems, this model can be used to describe rhythmic processes in general. The c/m and k/m parameters can represent abstractions of processes that result in oscillations and those which dissipate those oscillations, such as thermodynamic losses in chemical reaction networks. For those of you unfamiliar with the first law of thermodynamics (or simply need their memory jogged), this law states that the change in energy of a system, ΔE, results from the combination of work, w, done by or on the system and the heat, q, the system exchanges with the surroundings. No biochemical reaction is 100% efficient, i.e., one cannot simply equate $\Delta E = w$. So, there is always some loss in converting energy to work, which is dissipated as the heat, q. This fact explains why it is necessary to cool an automobile engine and why you sweat when you exercise. If a given process were 100% efficient, you would not need to compensate for the heat generated in the process. Since these processes are not 100% efficient—automobiles run at about 30% efficiency and skeletal muscle at about 45%—much of the potential energy available to any system is lost as heat. This is a major factor determining the design and function of both mechanical and biological devices.

From the perspective of the previous chapters, the biological models that relied on environmental cycles as drivers to sustain them result in damped oscillations when the environmental cycle is removed: they represent mathematical models that produce phase plane graphs similar to those on the right in Figure 6.3. Autorhythmic systems, such the autorhythmic neuron introduced in Chapter 4, can be represented by a mathematical model able to generate a phase plane graph similar to the one shown on the left side of Figure 6.3.

6.3   NONLINEAR MODELS OF OSCILLATORS

Spiral and elliptical orbits in the phase plane are the only oscillatory trajectories available to linear system models. Neither case is satisfactory for the description of a biological rhythm. Chemical reactions that drive rhythms in the body, like any real physical system, must obey the first law of thermodynamics. As such, dissipative forces have to be somehow included in the model. In terms of the linear equations, this means $c \neq 0$. At the same time, circadian rhythms are persistent processes that do not decay in amplitude. Unfortunately, to obtain such a result using the same linear equations, c must equal 0. Clearly, c cannot equal 0 and not equal 0 at the same time. How can we develop a mathematical model that resolves this paradox?

The linear dynamical systems described above only represent a particular kind of dynamical system model. More generally, dynamical systems are described by *nonlinear* systems that cannot be translated to a linear system as in Equation (6.3). Nonlinear systems can express a far richer diversity of behavior than linear systems, including multiple equilibrium points. Some of these equilibrium points may be stable and others may be unstable. Nonlinear systems can be analyzed based on the concept that within the *region* (or *neighborhood*) of an equilibrium point, a nonlinear dynamical system will behave like a linear system.

Mathematically, this represents a *linearization* of the nonlinear system, and it can be done through a relatively straightforward mathematical procedure based on finding the *Jacobian* of the system at each equilibrium point. If a solution trajectory in the phase plane approaches a stable equilibrium point, it will converge to that point and find a steady state. If the associated linear system in the neighborhood of that equilibrium point oscillates or spirals, then the nonlinear system will show oscillatory behavior. If an equilibrium point is unstable, then the solution trajectories will never converge to that point. A useful analogy might be to consider a seesaw. It naturally is at rest on one end or the other. If two children are on the seesaw and it is close to ground on a given end, it will tend to drop at that end. These two positions represent stable equilibrium points of the system. The children may be able to find a position that balances their moments such that the seesaw is suspended in the air. However, any small perturbation would cause it to fall down at one end or the other. This represents an unstable equilibrium point of the system.

Because a nonlinear system can sustain multiple equilibrium points and change its behavior in different regions of the phase plane, such system models can express more complex temporal behavior such as sudden switching, nonlinear oscillations, and chaos (fluctuations that appear to be effectively random).

Of particular relevance to the modeling of biological rhythm processes is a special kind of stable oscillatory trajectory that is available to nonlinear system models: the *limit cycle*. A classic example of a nonlinear model that results in a limit cycle is the Van der Pol model, described by the second order ODE:

$$y'' + a(1 - y^2)y' + y = 0 \tag{6.6}$$

This second order ODE can be rewritten as a system of first order ODEs using the same mathematical manipulation as for Equation (6.3) above:

$$
\begin{aligned}
y_1 &= a(y_1 - y_1^3/3 - y_2) \\
y_2 &= y_1/a
\end{aligned}
\tag{6.7}
$$

As an interesting sidebar for engineers interested in modeling biological systems: Balthazar Van der Pol was in fact an electrical engineer, and limit cycles were used to model vacuum tubes. However, his work was motivated by arrhythmias in the human heart[25]. Indeed, as Figure 6.4 shows, the phase portrait of the Van der Pol model shows that trajectories approach a stable circular orbit, representing the emergence of stable oscillations. Again, as in the example above, the actual form of the equation is obviously not anything close to an exact representation of any real biological system. It nevertheless still gives rise to similar qualitative behavior to what we would expect from a circadian rhythm and represents a starting point for more detailed modeling studies.

What makes the Van der Pol model behave differently from the linear model described above? It is the nonlinear damping term, which is represented by a quadratic term in Equation (6.6) and thus gives rise to the nonlinear y_1^3 term in Equation (6.7). Higher order Van der Pol models can be constructed with different damping terms, and those have in fact been implemented in models of circadian rhythms, e.g., by Kronauer in 1982, which was based on coupled Van der Pol models with a forcing function[17].

Like the linear model, this model has its equilibrium point at $y_1 = y_2 = 0$. However, this point is unstable, so the system cannot settle at that point in a steady state. Initial conditions near that point will lead to the system growing outward with oscillations that grow in amplitude exponentially. There is actually *negative damping*, because the damping term proportional to y' in Equation (6.6) will be negative, as opposed to positive as was the case in the previous mass-spring-damper model modeled by Equation (6.1). Thus, the phase portrait would show spirals outward, contrasting with the stable spirals shown in Figure 6.3. But when y in Equation (6.6) (or y_2 in the 2-D view of the system in Equation (6.7)) is large, the quadratic y^2 term on the damping becomes dominant. Thus, the damping will be positive again, and the oscillations become damped. Thus, the phase plane

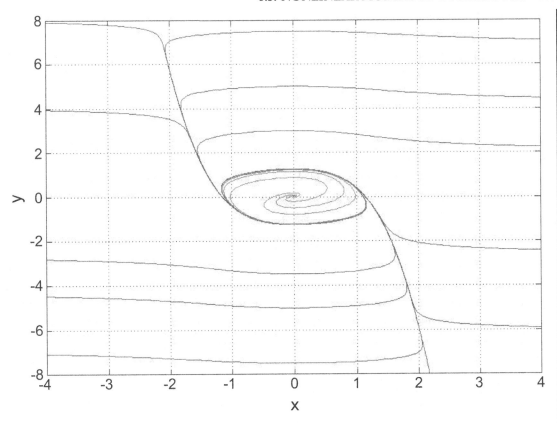

Figure 6.4: Limit cycle trajectories corresponding to $a = 1$ in Equation (6.7). Spiral trajectories that begin outside the limit cycle (blue) converge inward to the circle; trajectories inside the cycle (red) diverge outward until they reach the circle.

trajectories for initial conditions starting far away from the origin will tend to be spirals inward, as was the case shown in Figure 6.3.

This sets up a *bifurcation* in the Van der Pol model: when y is small, the phase behavior is unstable oscillatory, and when y is large, the phase behavior is stable oscillatory. As y gets smaller, the phase behavior would switch from stable to unstable oscillatory, but of course an unstable oscillatory behavior for small y would cause it to grow back up and become large enough that it would become stable oscillatory again. The resolution to this paradox is that the phase behavior will approach a circular orbit in the phase plane, corresponding to stable oscillations, as shown in Figure 6.4.

Notably for modeling perpetuating biological rhythm processes, limit cycles are remarkably robust. If a system is pushed off the limit cycle trajectory by a small perturbation, then the model predicts that it will return to the limit cycle without any need for additional interference. It can also be shown that if the model is adjusted by a small amount, a limit cycle will still exist. This is not the case for linear and nonlinear elliptical trajectories, like the circular trajectories in Figure 6.3, since any damping or dissipation would cause those to switch to oscillations with decaying amplitude.

In essence, this is the problem we encountered with our earlier models based upon the simple endocrine feedback control and environmental drivers. The resulting oscillations were too unstable to satisfy the twin requirements of internal and external synchrony. On the one hand, biological rhythms must be able to tolerate noise without losing temporal integrity, and on the other hand, they must be able to adapt to significant changes. Conceptually, we came to the conclusion that maintenance of stable internal oscillators would be the key evolutionary advance that would allow these requirements to become satisfied. We now know that there are mathematical and physical models that support the notion that these structures could have actually evolved biologically.

Circadian rhythm models, while more complicated than the elementary Van der Pol model shown here, generally predict the emergence of stable limit cycle trajectories. However, it is important to remember that these dynamical system models are simplified abstractions of the complex biochemical systems within cells. Systems of equations are constructed to represent key details of the system that the modeler wants to study. They are intentionally constructed to produce limit cycles under at least some ranges of parameters. If that did not happen, then the modeler would reject the system as being unrepresentative of the biological fact of robust oscillations. It is important to note that the *limit cycle* is not itself something that is physically real in some tangible way—it is just an emergent property of nonlinear dynamical system models that makes these models a practically useful tool to analyze rhythmic time series. Conversely, if a model were to be designed that could not simulate sustained oscillations, then it would clearly neither be an accurate nor useful description of systems whose oscillations we observe.

6.4   MODELING MOLECULAR NETWORKS IN CELLS

Models at different levels can be developed based on different abstractions of system variables. For example, the Van der Pol model can be expanded and modified to describe a physiological circadian pacemaker. The output variable might be the production level of a particular hormone in a physiological model. However, the details of the internal molecular network that generates that pacemaker in cells or groups of cells would be hidden in the parameters of the model. The parameters can only be fit to observed data. But changes in parameter values cannot in general be interpreted in terms of *specific* changes within cells and protein interactions, such as the loss of function of a particular gene.

For a model to be identifiable at that level, the variables need to represent the quantities of molecular species within biological rhythm systems. Then, conservation of mass principles can be used to generate a set of differential equations that balance the production and consumption of molecular species in chemical reactions. This is one of the most useful aspects of mathematical modeling. Otherwise, humans have great difficulty developing intuitive mental models that can conserve mass accurately. It is not hard to believe that it is the case, as anyone who tries to balance their checkbook in their head can attest! For those who want scientific evidence however, psychological studies of how people assess forces leading to climate change have shown this deficiency[24]. Thus, mathematical models are an invaluable tool to compare different hypotheses for molecular network structure and dynamics, especially for the complex systems that govern circadian rhythms.

Recently, the mathematical modeling of molecular networks in cells and tissues has gained a great deal of attention, as the data available from new technologies for microscopy, DNA sequencing (*genomics*), and protein measurement (*proteomics*) among others have become more available. However, the analysis of molecular networks involving mRNA transcription from genes, protein production, and enzymatic reactions predates these new technologies. In fact, in 1963, Brian Goodwin published *Temporal Organization in Cells*[11], in which he described fundamental equations that can describe interacting networks of mRNA and proteins in cellular processes. Readers interested in biological system modeling are encouraged to consult research from that era in addition to the substantial literature available today.

Goodwin developed a model of a transcription-translation loop, wherein a protein inhibits the production of mRNA from its encoding gene[12]. He showed that the loop could exhibit oscillations under certain parameter values. The early models that he analyzed showed large and unstable oscillations, but he was able to show that models of coupled oscillators can show pulse entrainment behavior and other results that agreed with observed behavior of circadian rhythms. To illustrate how these negative feedback loops behave in practice, let us consider the two activator-inhibitor models shown Figure 6.5. Using conservation of mass for the concentration of X and Y protein molecules,

Figure 6.5: Models of activator-inhibitor feedback loops. We can consider two proteins, an activator and an inhibitor protein (X and Y respectively). In Model A, on the left, the promoter for transcription of X mRNA is inactivated by the binding of a single Y protein, and the promoter for transcription of Y mRNA is activated by the binding of a single X protein. Model B, on the right, includes cooperativity between X proteins, in which 2 proteins of X activate the X promoter.

represented by x and y respectively, we can write down the following set of equations:

$$\text{Model A:} \quad \begin{aligned} \frac{dx}{dt} &= v_x + k_x \frac{A_1}{A_1 + y} - \gamma_x x \\ \frac{dy}{dt} &= v_y + k_y \frac{x}{A_2 + x} - \gamma_y y \end{aligned} \tag{6.8a}$$

$$\text{Model B:} \quad \begin{aligned} \frac{dx}{dt} &= v_x + k_x \frac{x^2}{A_2^2 + x^2} \frac{A_1}{A_1 + y} - \gamma_x x \\ \frac{dy}{dt} &= v_y + k_y \frac{x}{A_2 + x} - \gamma_y y \end{aligned} \tag{6.8b}$$

Here, we can identify v_x as the basal rate of X protein production (i.e., a basal mRNA transcription rate), v_y is the basal rate of Y protein production, γ_x and γ_y are decay rates of X and Y proteins respectively, and k_x and A_1, k_y and A_2, are Michaelis-Menten kinetic parameters for X and Y production under activated conditions. Model B has been modified to introduce *cooperativity*, i.e., by assuming that X would be activated by the cooperative action of two X molecules, as show in Figure 6.5. Mathematically, the cooperativity term in Model B is provided in (6.8b) by squaring the terms in the dx/dt equation, which takes on a form known as a *Hill model*. This power of 2 is generally referred to as a *Hill coefficient* of 2. The inhibition of X by Y is included with the $[A_1/(A_1 + y)]$ term in both models, which is constructed so that the fraction does not approach infinity when the level of Y is low (and y is consequently small).

The Equations (6.8a) and (6.8b) can be simplified by assuming that $A_2 \gg x$. (In words, the production of X will not approach its maximal transcription rate). In addition, we can rescale some the variables and parameters, leading to the following reduced model:

$$\text{Model A:} \quad \begin{aligned} \frac{dx}{dt} &= \gamma_x \left(v_x + k_x \frac{1}{1 + y} - x \right) \\ \frac{dy}{dt} &= v_y + k_y x - y \end{aligned} \tag{6.9a}$$

$$\text{Model B:} \quad \begin{aligned} \frac{dx}{dt} &= \gamma_x \left(v_x + k_x \frac{x^2}{1 + x^2} \frac{1}{1 + y} - x \right) \\ \frac{dy}{dt} &= v_y + k_y x - y \end{aligned} \tag{6.9b}$$

The variables and parameters shown here used the same symbols as in Equations (6.8a) and (6.8b) for clarity. It can be shown that the equations for Model (6.9a) lead to a stable steady state in the positive $x - y$ quadrant of the phase plane. This is what happens in the physically realistic situation, since x and y must be positive. These phase trajectories are illustrated for a sample set of parameter values in Figure 6.5. As Figure 6.5 shows, the equations for Model (6.9b) can lead to a stable limit cycle in the positive quadrant (for values of $\gamma_x > 1$). Regardless of initial conditions, the system will converge to that limit cycle and oscillate stably.

We can see that to obtain this behavior there is a need for more complexity in the equation than the simplest model. In this case, this complexity comes from replacing the Michaelis-Menten formulation with a Hill formulation, i.e., introducing the quadratic term x^2 in (6.9b). Other work has shown that adding additional variables to the model and increasing can also lead to stable limit cycle behavior. Note that increasing the number of variables will also increase the number of equations in the system, making it harder to analyze with 2-D phase portraits as shown in Figure 6.6. Thus, modeling papers will often show multiple phase portraits with different combinations of variables plotted against each other.

We will now take this concept of a negative feedback loop model further and describe an example of its application to modeling a pacemaker process in the cell, as outlined in Lema et al[18]. In their model, the negative feedback equations include a delay term:

$$\frac{dP(t)}{dt} = K_e G(t - \delta) - K_d P(t) \tag{6.10}$$

$$G(t - \delta) = \frac{1}{1 + [P(t + \delta)/K_i]^n} \tag{6.11}$$

In these equations, $P(t)$ is the amount of protein at time t, and $G(t)$ is the "activity" of the gene encoding the protein, which represents an abstraction of the transcription rate. The protein is assumed to decay linearly with rate K_d. The protein inhibits gene activity, modeled by a Hill model with inhibition parameter K_i and a Hill coefficient of n representing cooperativity between multiple proteins. These equations include a delay term δ: the protein is produced at a rate proportional to the gene activity at δ time units previously. Based on these models, Lema and coworkers[18] showed that to achieve oscillations at the 24-hour period of the circadian rhythm, they had to assume a delay time δ of approximately 8 hours.

The effect of lag times in determining the period of oscillations was initially introduced in Chapter 1 with our discussion of room temperature and thermostats (Chapter 1, Figure 1.1). Although considerably more sophisticated, these equations illustrate the important conceptual point made in Chapter 1 that different lag times can generate oscillations with different periods using the same basic components.

However, using a similar negative feedback loop model with a delay, another group[22] predicted a delay of 4 hours. In their case, they explicitly modeled the production of mRNA with its own differential equation, rather than assuming an algebraic equation for an abstracted "gene activity" as Lema and colleagues[18] did. Letting the time-dependent concentration of mRNA be $M(t)$ and assuming that mRNA decays linearly at a rate K_{d2},

$$\frac{dP(t)}{dt} = K_e[M(t - \delta)]^m - K_{d1} P(t) \tag{6.12}$$

$$\frac{dM(t - \delta)}{dt} = \frac{1}{1 + [P(t - \delta)/K_i]^n} - K_{d2} M(t - \delta) \tag{6.13}$$

In Equation (6.12), it is assumed that the production of protein depends on mRNA concentration through a power law. The Hill coefficient m in this case represents nonlinear aspects of

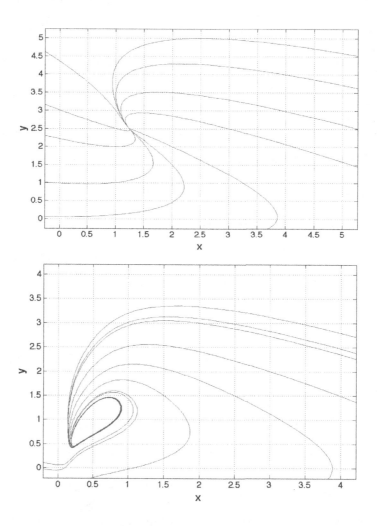

Figure 6.6: Phase portraits of the negative feedback loop models shown in Figure 6.5 and described by Equations (6.9a) and (6.9b). Phase trajectories for *Model A* are shown on the top converging to a stable steady state. *Model B*, on the bottom, shows convergence to a stable limit cycle (the heavy lines at the lower-left of the plot). The parameter values used to generate these plots were $v_x = 0.1$, $v_y = 0$, $k_x = 4$, $k_y = 2$, and $\gamma_x t = 5$ (these are the rescaled parameters in Equations (6.9a) and (6.9b)).

the production of protein from mRNA. For example, different mRNA may be transcribed to form protein components that come together to form the final clock protein. Thus, although the system of Equations (6.12)–(6.13) explicitly includes a variable called mRNA concentration, it is essentially a model at the same level of abstraction as the system of Equations (6.10)–(6.11).

The fact that the delay term δ has to take on a significantly different value to predict a limit cycle with a 24-hour oscillation period reveals two important insights. First, the structure of the equations makes a significant difference in model predictions. Second, the abstractions on the equations are concealing biochemical details that could be used to make more accurate predictions and discriminate between divergent models of cellular rhythm generating processes.

It is interesting to note how well Goodwin's modeling work[11,12] anticipated progress in elucidating the actual molecular mechanisms of biological clocks. Eight years later, three mutations were identified which significantly altered *Drosophila* circadian pupal eclosion (emergence from the pupa stage) rhythms[16]. Further investigations into the genes responsible and their protein products eventually lead to the development of the transcription feedback models of eukaryote circadian rhythms[3,13,14,15,26,27] demonstrating how mathematical models can presage the discovery of the mechanisms underlying the systems being modeled.

6.5 MODELING EXTERNAL PERTURBATIONS ON BIOLOGICAL OSCILLATORS: SYNCHRONIZATION, ENTRAINMENT, AND OTHER EFFECTS ON RHYTHMS

The models in the previous section describe the behavior of individual oscillators. However, in tissue such as the brain and heart, multiple cells express physiological rhythms through synchronized oscillations of their rhythm-generating molecular networks. Furthermore, these rhythms are entrained with particular period lengths defined by exogenous oscillations, such as the light/dark cycle. In this section, we will briefly discuss approaches to studying these phenomena with nonlinear dynamical system models, focusing on providing the background to the terminology used in the modeling literature.

To study the synchronization of coupled oscillators, Gonze and colleagues[10] modified the Goodwin oscillator to include the activity of a clock gene mRNA (x) that produces a clock protein (y) that activates an inhibitor (z) of the transcription of the clock gene:

$$\begin{aligned}
\frac{dx}{dt} &= v_1 \frac{k_1^n}{k_1^n + z^n} - v_2 \frac{x}{k_2 + x} \\
\frac{dy}{dt} &= k_3 x - v_4 \frac{y}{k_4 + y} \\
\frac{dz}{dt} &= k_5 y - v_6 \frac{z}{k_6 + z}
\end{aligned} \tag{6.14}$$

As always, it is important to remember that Equation (6.14) represents a highly abstracted model for the much more complex molecular networks that actually control rhythm processes in cells, and the constants $k_1 - k_6$ and $v_1 - v_6$ represent lumped parameters. Michaelis-Menten formulations are used for all the processes except transcriptional inhibition, which is represented by a Hill function (representing possible cooperativity between multiple proteins to inhibit gene transcription). When the Hill coefficient is set to $n = 4$, the system gives rise to limit cycle oscillations. To represent the coupling between oscillators and the entrainment by an exogenous oscillation, the model is modified as follows:

$$\frac{dx_i}{dt} = v_1 \frac{k_i^n}{k_i^n + z_i^n} - v_2 \frac{x_i}{k_2 + x_i} + v_c \frac{KF}{K_c + KF} + L$$
$$\frac{dy_i}{dt} = k_3 x_i - v_4 \frac{y_i}{k_4 + y_i} \tag{6.15}$$
$$\frac{dz_i}{dt} = k_5 y_i - v_6 \frac{z_i}{k_6 + z_i}$$

There are now N oscillating cells $i = 1, 2, \ldots, N$. The external signal is represented by the function of $L(t)$ that can be a sinusoidal or square waveform representing a light/dark cycle or other entraining signal (*Zeitgeber*). In the Gonze model, the cells were coupled together by assuming that the mean level of a neurotransmitter released by cells, represented in Equation (6.15) by $F(t)$, has an effect on the transcription of the clock gene x_i. In their paper, Gonze and colleagues showed how varying the strength of the coupling between oscillators, represented by the parameter K in Equation (6.15), can cause the overall system of N cells to generate sustained limit cycle oscillations, damped oscillations, or convergence to a steady state level.

In the absence of coupling, these models can show some rather interesting features. For example, models similar to Equation (6.15) for single or multiple oscillators can generate either sustained limit cycles that eventually converge to a period near that of L or chaos for various amplitudes of light/dark oscillations in the function $L(t)$. In other words, the models can display either entrainment to L or arrhythmicity depending on the interactions between L and the oscillator(s). Chaos arises through a series of bifurcations in the limit cycle. Essentially, the period of the limit cycle keeps doubling, and as this happens the oscillations change amplitude and frequency such that the resulting temporal patterns become indistinguishable from noise. When light/dark oscillations are sinusoidal, the system model tends to generate stable oscillations. But as the light/dark temporal patterns look more like square waves, the system model will lead to chaotic behavior for increasingly large ranges of parameter values. Finally, in the case of square waves, chaos is quite likely for many parameter values. Analogous phenomena have been described for other systems, such as electronic circuits.

The explanation goes back to the Fourier transform representation of temporal waveforms. Any periodic wave can be decomposed into a sum of sinusoidal waves of different frequencies or *harmonics* which are integral multiples of the fundamental frequency of the wave. A sinusoidal wave represents a single harmonic. A square wave is made up of an infinite number of harmonics, as shown in Figure 6.7. While the higher harmonics have much smaller amplitudes than that of the fundamental frequency, they do not decay to values approaching zero as they do for more "curving" waveforms. Thus, there are many more frequencies to which the system will try to entrain itself. The bifurcations that lead to chaos will be much likelier to result in this case. This phenomenon has been studied in simulations[9]. The authors of that work proposed that physiological circadian rhythm systems evolved to adapt to light/dark cycles that are not sudden transitions. Thus, they will behave most stably when there are more gradual and continuous increases and decreases in light levels during the cycle, which contrasts with the conditions used in many animal experiments.

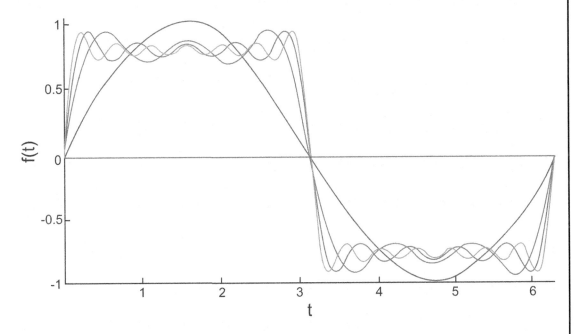

Figure 6.7: Fourier series for a square waveform. Shown here is a single sine wave (the fundamental) and then the sum of successive odd harmonics (i.e., sine waves with odd multiples of the frequency of the fundamental sine wave) showing how the sum approaches a square wave as more frequencies are summed up.

This may explain the impact of night shift and jet travel on human health and behavior as discussed in Chapter 5. Multiple shifts in the external environmental cycles may be viewed as multiple cycles by the circadian pacemakers and oscillators that make up the human circadian system. Under such circumstances, the circadian system would have a tendency to display chaotic behavior and lose both internal and external synchrony.

When we introduced the phase plane representation as shown in Figures 6.3 and 6.4, it was motivated by the desire to show the qualitative behavior of dynamical systems instead of the quantitative changes in the level of system variables over time. The latter depends on the initial state of the system, and it may not be significant for system function. For example, the quantitative level of a clock protein within a cell may not be biologically significant, or even if it is, may not be measurable to allow model validation and accurate parameter estimation. However, showing that the clock protein level stably oscillates is highly significant to the study of the rhythm process.

Similarly, when studying the synchronization and entrainment of circadian oscillators, modelers are generally more interested in oscillation periods of the mRNA and protein components of the system than their amplitudes. In particular, they are interested in the phase differences between the oscillations of the different components and the entraining signal, such as the light/dark oscillation period. Amplitude *changes* are not significant since following a perturbation, a dynamical system that expresses limit cycle behavior may change amplitude but eventually the amplitude will settle back at its original level because the limit cycle is itself stable (as illustrated for example in Figure 6.4).

A key observation is that the effect of perturbing the phase of a variable is highly dependent on the phase at which the perturbation occurs within the cycle of the variable. For example, for human circadian clocks, light pulses given in the morning advance circadian phase but when given at midnight, they delay the phase. To study this phenomenon biological rhythms often use what is called a *phase response curve* representation (often abbreviated to PRC), a schematic of which is illustrated in Figure 6.8a. This is a plot of the phase change (Δf) observed in a variable in response to a pulse given at an original phase f. As shown in Figure 6.8, there will be critical points at which the phase difference Δf will be zero. The related *phase transition curve* is shown in Figure 6.8b. It shows the new phase that arises from a given original phase in response to a perturbation. Recall that we were able to generate a crude phase response curve when discussing the effects of pulsed stimulation on the frequency and phase of an autorhythmic neuron in Chapter 4. That curve is recreated in Figure 6.9.

The effect of multiple perturbations can be described using an iterated map, which can be represented by the following equation:

$$\phi_{n+1} = f(\phi_n) \tag{6.16}$$

What this means is that the new phase (ϕ_{n+1}) following the nth perturbation is calculated based upon the previous phase (ϕ_n) though a function (f) that represents the transition between phases. An example for such a function would be the system response to the exogenous signal within Equation (6.15). Equation (6.16), in fact, represents a form of a dynamical system and it has equilibrium points which are the f_c shown in Figure 6.8. Starting at some initial phase f_0, subsequent steps of updating the phase will end up having smaller and smaller differences until you effectively converge to a final value. Some of these equilibrium points will be stable and others will be unstable. The condition for stability is that the slope of the phase transition curve be small, i.e., $|df/df(f_c)| < 1$. Steeper slopes (i.e., derivative greater than one) indicate an unstable equilibrium point.

Note: A simple example of an iterated map is $x_{n+1} = x_n^2$. If $x_0 > 1$, there is no convergence; the sequence simply increases toward infinity. If $|x_0| \leq 1$, then there are two equilibrium points at $x = 0$ and $x = 1$. The first equilibrium point is stable. If you start computing the iterated map in this case with any number less than or equal to 1 other than 1, squaring the result repeatedly will lead to

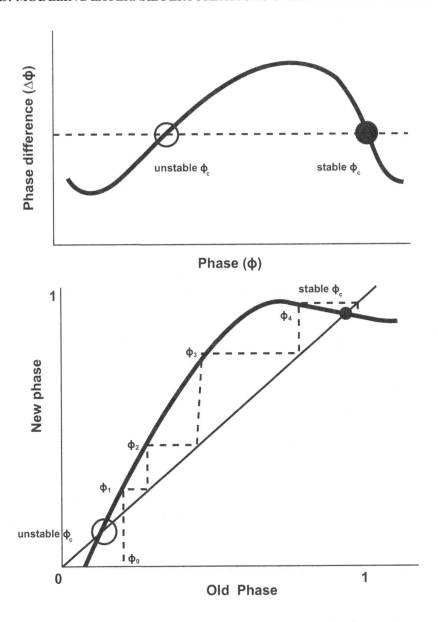

Figure 6.8: Example of a phase response curve (above) for a system showing stable (filled circle) and unstable (empty circle) equilibrium points. The phase transition curve below shows how the stable equilibrium phase is reached after a series of iterations (represented as dashed lines) from any initial phase (f_0).

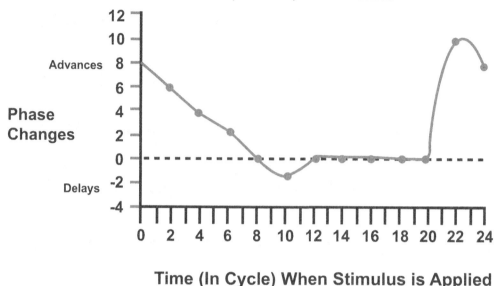

Figure 6.9: Phase change vs. time of stimulus application graph for hypothetical autorhythmic neuron. (See Chapter 4 for full explanation.)

0, e.g., $x_0 = 0.5$, $x_1 = 0.5^2 = 0.25$, $x_2 = 0.25^2 = 0.0625$, $x_3 = 0.0625^2 = 0.0039$, etc., becoming smaller with each step.

The slope of its phase response curve can allow us to predict whether a rhythmic system can be entrained. Entrainment requires a phase change: for example, if the endogenous period of the system is 23 hours and the exogenous cycle is 24 hours, the phase must be stably delayed by 1 hour. The vertical axis of the phase response curve shows a phase change that can be corresponded back to an original phase on the horizontal axis. If the phase response is stable at that point (i.e., negative slope) then stable entrainment is possible. Notably, the amplitude of the entraining signal relative to the endogenous rhythm *is* significant here, because it can change the shape of the phase response curve, allowing for a greater range of stable entrainment periods.

The sample phase transition curve in Figure 6.8 shows the phases increasing as an iterative map along a line that indicates 1:1 entrainment of frequencies. The iterative map of phase transitions can be used to analyze frequency locking behavior observed in coupled oscillators. In 1673, Christian Huygens observed that pendulum clocks are coupled through oscillations due to vibrations in the wall. This led to a concept that when oscillators are coupled even weakly, synchronization can occur if the frequencies are close enough. As the coupling strength increases, the deviation between frequencies can increase even more and still allow for frequency locking. The locking occurs at a rational fraction between frequencies, such as 1:1 or 2:3. Figure 6.10 shows an example of frequency locking as a function of forcing strength or amplitude and the coupling between oscillating signals.

It shows the behavior for a simple iterated sine map,

$$\phi_{n+1} = \phi_n + \Omega + k \sin(2\pi \phi_n) \tag{6.17}$$

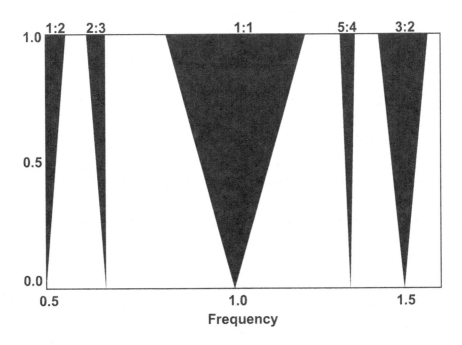

Figure 6.10: "Arnold tongues" entrainment at p:q rational ratios for the sine map. Similar patterns are generally found for coupled limit cycle models.

Here, Ω is a parameter related to the frequency ratio and k is the coupling strength of the oscillation. The regions at which the frequencies lock at a particular ratio are indicated as dark areas in the graph of Figure 6.10. These can be interpreted as *resonance zones* or *entrainment zones*. They are sometimes referred to as *Arnold tongues* in the literature, after Vladimir Arnold, a Russian mathematician. Similar regions are found for more complex iterative maps, such as those arising from the Gonze models of Equation (6.15).

An important issue in the analysis of circadian rhythm models is the observation that certain kinds of perturbations—such as a constant, high intensity stimulus—can cause rhythms to stop. Within the context of modeling, this is sometimes called *singularity behavior*, and it has been related to the limit cycle model of biological rhythm systems. If we go back to the limit cycle portrayed in Figure 6.4, there is an unstable equilibrium point at the origin (where the interior spirals colored in red emerge to join the limit cycle). In the singularity behavior view of this phenomenon, certain perturbations—such as constant high stimulation—can cause the system to be held at the unstable equilibrium point at the center of the limit cycle that governs system behavior. This then results in arrhythmicity. A more complex explanation that can account for the persistence of arrhythmicity is based on the view that rhythms result from multiple synchronized oscillators. The perturbation still brings each individual system to the unstable equilibrium point—and when released, each individual system returns back to the limit cycle. However, since the phase of the limit cycle for each oscillating system is now essentially randomized relative to all the other oscillators, the average behavior of all the systems together cancels each other out, resulting in sustained arrhythmic behavior for the system as a whole.

6.6   PHASIC ENTRAINMENT, PARAMETRIC EFFECTS, AND RHYTHM POSITION

We will now discuss the effect of changing the ratio between the period length of the entraining oscillator (which we will label $\mathbf{T}$) and the period of the entrained oscillator (τ or tau) on the phase angle between the two rhythms (β). To do so, we turn back to the linear models described previously. While a more complex model of coupled nonlinear oscillators might more accurately describe the relationship for real biological systems, the mathematics is significantly more difficult. Moreover, coupled linear oscillators and coupled nonlinear oscillators have roughly similar qualitative behavior. Assuming a linear model, we can write down equations for a synchronized (entrained) oscillator as follows, representing the entrained oscillator as a mass-damper-spring system, just as we did earlier in the chapter. We start with the basic form of Equation (6.2) for the displacement of the mass (y), and on the right-hand side we include a sinusoidal function representing the entraining oscillator, with amplitude A_0 and period $\mathbf{T}$ driving the entrained oscillator described by the terms on the left-hand side.

$$y'' + (c/m)y' + (k/m)y = A_0 \sin\left(\frac{2\pi}{T}t\right) \tag{6.18}$$

As we discussed in the context of Equations (6.1)–(6.5), in the absence of the driving term due to the entraining oscillator, the solution to Equation (6.18) will involve an exponential rate depending on the (c/m) term and oscillatory frequency that depend on the (k/m) term. Since we are interested in the phase angle as a function of the periods of the oscillators, we will write Equation (6.18) in a modified form:

$$y'' + 2\gamma y' + \omega^2 y = A_0 \sin(\omega_0 t) \tag{6.19}$$

Assuming that the qualitative behavior of the entrained oscillator will be in the form of damped oscillations based on the relationship between the spring and damper parameters, as we discussed in the context of Equation (6.5), then the solution of (6.19) can be written down as:

$$y(t) = \exp(-\gamma t)(A\cos(\omega t) + B\sin(\omega t)) + \frac{A_0 \sin(\omega_0 t + \beta)}{\sqrt{(\omega^2 - \omega_0^2)^2 + (2\gamma\omega_0)^2}} \tag{6.20}$$

where A and B are constants whose values are specified by the initial conditions and damping parameters. Note that the first of the two terms summed together in (6.20) represents the damped sinusoidal solution of Equation (6.2). This term only describes the transient behavior of the system immediately after it starts oscillating, since as time progresses, the exponential component will decay to zero. What will be left is continuing oscillation with the amplitude as described in the second term in Equation (6.20), at the same frequency as the entraining oscillator but with a phase shift of β, where:

$$\beta = -\arctan\left(\frac{2\gamma\omega_0}{\omega^2 - \omega_0^2}\right) \tag{6.21}$$

We can write this equation for the phase shift of the entrained oscillator with respect to the entraining oscillator as depending on their periods rather than frequencies. We identify the period of the entraining oscillator as $T = 2\pi/\omega_0$, and since the damped oscillation term in Equation (6.20) represents the inherent behavior of the entrained oscillator in the absence of any driving, we identify its period as $\tau = 2\pi/\omega$.

We can plot the phase shift β as a function of some values for the ratio T/τ. What we find agrees with what may be expected from empirical observation. As the ratio decreases, i.e., the period length of the entraining oscillator T increases or τ gets shorter, the phase angle β increases, and the phase of the entrained oscillation advances relative to the entraining oscillation. Note that where the period lengths of the two oscillators are the same there is a singularity in Equation (6.21). This is reflected in Figure 6.11 as the point where the phase angle changes sign.

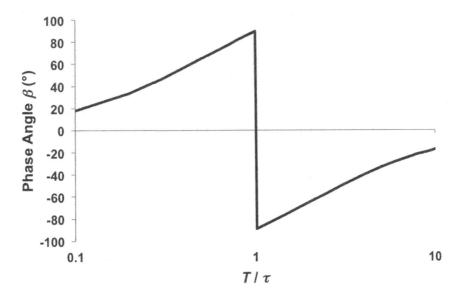

Figure 6.11: Change in the phase angle β for varying ratios of entrained and entraining oscillator period lengths T/τ. The values here are plotted for an arbitrarily selected value of $\gamma = 10$, and the ratios are plotted on a logarithmic scale so that it is easier to see the overall trend.

It is important to note that this analysis involves a *linear* model. As a result, our work excludes a lot of the details of how entrainment occurs in the nonlinear context. Nevertheless, at the narrow limit where the linear driven oscillator model does describe the qualitative behavior of nonlinear models, the basic relationship between the period ratio and phase angle will be the same.

What are the practical implications for this phenomenon in the case of biological rhythms? Our previous discussions have focused on the use of phase changes to generate an apparent synchronization between two oscillators. In Chapter 4, a pulsed signal (source unknown) was used to synchronize an autorhythmic neuron (see Figure 6.9) using a phase response curve. In this chapter, we extended this crude model with greater mathematical precision to show how a geophysical cycle (Zeitgeber) such as the day/night cycle could synchronize or entrain a circarhythm such as a circadian cycle using a similar approach (i.e., a phase-response curve; see Figure 6.7). Theoretically, the process of so-called *phasic entrainment* does not require any alteration in the fundamental period of the entrained oscillator. This leads to the possibility of three important periods—that of the entraining oscillation (T), the period of the underlying mechanism creating the entrained oscillation (τ) and the apparent period of the entrained oscillator which results from the interaction of T and $t - \tau*$ or *tau apparent*.

An example should make this idea more clear to you. Suppose you owned a watch that ran 5 minutes fast every 24 hours. Realizing this, and being so frugal that you do not wish to repair it or buy a new one, you reset the watch by 5 minutes every morning at 7:00am when you go to work using the wall clock as a reference. Assuming the wall clock is completely accurate, you are now dealing with three periods—the period of the wall clock (T = 24 hours), the period of the watch itself (τ = 23 hr, 55 min.) and the corrected or apparent period of the watch (τ = 24 hours). You, incidentally, represent the coupling mechanism linking the cycle of the wall clock to the oscillator in the watch.

Thus, we can see that maintaining synchronization between two cycles using phase changes can position a rhythm properly within an entraining cycle without any significant alteration of the fundamental period of the entrained oscillator. Analogously, phasic entrainment should work to position circadian rhythms within the geophysical cycle of day and night without having to change the period of the underlying circadian clock. In fact, the same should be true for any set of entrain*ing* and entrain*ed* biological oscillators.

However, our original autorhythmic neuron model displayed another characteristic in response to input stimuli—a change in the period or angular velocity of the neuron's oscillatory behavior could be generated in response in different levels of constant stimulation. Potentially, this effect could be seen with other oscillators, including circadian clocks, if a mechanism existed to shift period in response to stimulus intensity, for example, by changing the lag times. This leads to the interesting possibility that environmental stimuli could have multiple effects on rhythmic phenomena by exerting both phasic and period-changing (called *parametric*) effects simultaneously.

Consider the following possibility. A nocturnal (night-active) animal is entrained by the geophysical cycle of day and night. According to our mathematical models, that could be achieved by repeated phase changes without any alteration in the animal's underlying circadian period. The result would be a frequency match between the day/night cycle and the circadian clock and a stable phase angle between the two rhythms. However, that does not seem to be quite enough to ensure an ecologically appropriate timing. After all, it is not sufficient if the animal merely has a peak activity at the same time every day—it must be active at the most favorable time every day. For a nocturnal animal, clearly, that is at night. Unfortunately, that is more of a problem than it at first appears, because in the real world of seasons, night is not always the same.

Recall the seasonal changes in photoperiod discussed in Chapter 3 (see Table 3.1). For example, Philadelphia's sunset occurs around 4:30pm in December and around 8:30pm in June—a 3 hour difference (taking in account the effect of Daylight Savings Time). If we assume that our hypothetical nocturnal animal were to begin activity at 5:30pm, it would be appropriately dark in December but broad daylight in June. Clearly, the animal must adjust the phase of its activity cycle to compensate for the seasons. How could this adjustment evolve?

The mathematical modeling of entrainment and the T/τ cycle effects on phase provide one plausible answer. If increasing amounts of light (or decreasing duration of darkness) were to alter the animal's underlying period (τ), it would be possible to alter the phase positioning of the activity rhythm without losing entrainment. For example, if greater amounts of light (available in the summer months) were to lengthen τ, then according to the T/τ relationship, the phase of the entrained circadian clock (and the activity it controls) would be delayed in the summer, that is, occur later. In the winter, when light decreased (or darkness increased), the period would shorten and the phase would advance, i.e., activity would occur relatively earlier. Despite the fact that light (or dark) intensity is changing τ, phasic entrainment remains intact since that requires merely that $T = \tau*$ and not that $T = \tau$. Again, we see how the mathematics of oscillators provides a potential evolutionary pathway for the further development and refinement of biological rhythms.

Incidentally, what we describe as an effect of Zeitgeber stimulation on circadian period is not merely a hypothetical exercise. Many nocturnal animals display a longer period associated with greater light intensities, exactly as the modeling predicts[1], possibly as a mechanism to generate the best temporal positioning for seasonal adjustment.

6.7   MORE DETAILED MOLECULAR NETWORK MODELS

The molecular network that generates a pacemaker signal in cells is clearly more complex than a single mRNA-protein or activator-inhibitor feedback loop as described in the previous section. Indeed, quantitative data are available for levels of mRNA and protein for circadian rhythms of several clock proteins, and those can be included in more detailed biochemical data. Most contemporary modeling work involves the generation of these more complex models, with more variables and more differential equations.

However, there is still a place for models at a higher level of abstraction, like the negative feedback loop models described in equations above and even variations of the simplified Van der Pol model. When developing physiological models of circadian rhythms, the contribution of pacemaker cells may not have to be explicitly modeled with biochemical details. What matters in that case is that there is a signal being generated with an appropriate limit cycle behavior that agrees with experimental observations. Including additional biochemical details at the level of intracellular processes would be superfluous, since the model would be used to predict hormone levels and other measured physiological parameters at a much higher scale.

With the advance of automated molecular biology technologies such as genomics and proteomics, increasing amounts of molecular data are becoming available that complement physiological data sets. One might, for example, ask what effect a mutation in a particular clock gene might have on the physiological response of an animal. In that case, it may be important to embed a more detailed biochemical detail within a broader physiological model designed at a higher level of abstraction. These *multiscale* modeling approaches are become more significant for a wide array of physiological systems, such as modeling the heart. (See, for example the Physiome project, www.physiome.org.)

One example of a highly detailed biochemical model is that developed by Forger and Peskin[4]. They modeled clock protein interactions in the suprachiasmatic nucleus (SCN) cells of the mammalian brain. Their model was based largely on circadian data for the mouse obtained by Reppert and Weaver[21]. The modeled system incorporates the activity of PER1 and PER2 proteins that transport the CRYPTOCHROME proteins CRY1 and CRY2 into the nucleus, where they inhibit production of the PER and CRY proteins. Thus, at the basis of the model is a negative feedback loop (CRY inhibits CRY) essentially like the models shown in Figure 6.5 and described by Equations (6.10)–(6.11).

However, Forger and Peskin[4] included many more interactions than the simple activator-inhibitor feedback models. For example, as an additional set of variables, they include the activity of REV-ERBa explicitly, the transcription factor that actually binds to inhibit production of the CRY proteins. They explicitly modeled the translation of PER1, PER2, CRY1, and CRY2 proteins; binding and unbinding of PER proteins with kinases; the phosphorylation reaction of kinases with the PER proteins (a required step for them to be activated); the binding of REV-ERBa to transcription factor binding sites in the promoters of CRY1 and CRY2 genes; the degradation of all the protein and mRNA species; along with many other reactions not mentioned here. In total, they define more than 30 reaction classes. The model includes a total of 74 variables governed by 73 differential equations (the discrepancy is due to their assumption that total kinase concentration is constant in time).

As is generally the case for such molecular system-scale models, the differential equations in the Forger and Peskin model[4] are defined based on conservation of mass between the variables. The assumption generally is that reaction kinetics follow the law of mass-action. That is, the reaction rate is proportional to the concentration of the reactants, with power law coefficients determined by the stoichiometry of the reactions. The only exceptions are the assumption of cooperativity in transcriptional activation. Even though the model only includes the activity of 5 constituents of the pacemaker system, it still results in a system of differential equations with a large number of parameters (each reaction class has an associate reaction rate constant, and in the case of cooperative reactions, more than one). This demonstrates how even models of a limited part of a biological system can explode in size as the depth of biochemical detail increases.

Biochemical experiments to accurately measure particular reaction rates *in vitro* are time-consuming and can generally not be done in a high-throughput format, unlike the measurement of protein and mRNA levels for the whole system. Furthermore, a measurement of reaction rates *in vitro* under controlled conditions is often a poor representation of what actually goes on in a cell, where many co-factors may transiently participate in reactions through mechanisms that are not well understood.

Consequently, most parameters have to be inferred from experimental data obtained from measurements of system variables other than the kinetics of the reactions themselves. The approach used by Forger and Peskin[4] was to first determine a likely set of parameters through trial and error, under the assumption that their model had to predict a circadian rhythm oscillation. Then, the parameter values of the model were refined through an iterative to obtain the best least square fit to the experimental data set reported by Reppert and Weaver[21]. Thus, the result that their model predicts a circadian rhythm with appropriate periodicity and relationships between CRY and PER protein and mRNA dynamics is unremarkable, since those may be simply a function of fitting a complex nonlinear model to data. As an alternative strategy for parameter estimation in rhythm models, Bagheri and coworkers[2] proposed fitting to phase behavior rather than amplitudes of signals over time, which takes advantage of the relatively greater stability in the phase behavior of limit cycles.

Notably, even when a great deal of data is available, nonlinear models can be susceptible to overfitting, unlike the case for linear models. This is because phase trajectories for highly dimensional (i.e., lots of variables) nonlinear models may be identical to those generated by a different lower dimensional nonlinear model with few variables. Thus, many different large nonlinear models may predict even a large experimental data set with equivalent precision. This makes it difficult to discern which model contains the biologically correct reaction mechanisms. Thus, the process of validating a biological system model is generally quite complicated. In the case of the Forger and Peskin[4] model, they had available data for the circadian rhythm generated by PER and CRY single and double mutants for the SCN as reported by Reppert and Weaver[21].

This is a somewhat crude test, since it involves fully knocking out PER and CRY activity from the model. Ideally, a quantitative model would be tested by seeing if it can dynamically respond to different gene doses of PER and CRY, but that can only be done very roughly given currently available genetic manipulation techniques. The Forger and Peskin[4] model was able to show good agreement with physiological data, at least in terms of predicting the ability of mutants to produce circadian rhythms, giving some confidence in model predictions.

Besides the problems of parameter estimation and model validation, detailed biochemical models also suffer from challenges in analysis. In the latter case, there is still a reasonably manageable number of model components. It is possible to interrogate the model manually for different permutations of PER and CRY gene deletion, and determine the relative significance of different aspects of the model in producing oscillatory behavior. However, one can imagine that as the number of proteins, genes, and other chemical species (e.g., calcium, ATP, etc.) increases, analyzing and visualizing model predictions becomes a computationally intensive task. It is possible to generate plots using computer simulation for different parameter ranges and equation configurations to analyze stability characteristics, the ability to robustly generate limit cycle behavior, and other aspects of the model. But for larger number of variables in complex models needed for biological realism, there is an outstanding need for new approaches to visualize and analyze results with the ease that is possible for two- and three-dimensional systems. This may include projecting high-dimensional models onto lower numbers of dimensions, or studying different components of models individually (such as is done by SPICE for electrical circuits).

As a final note, all of the discussion so far has been of deterministic, continuum models using differential equations. There are two important caveats for applying such an approach at the cellular level. First, there can be substantial intrinsic noise in biochemical reaction networks due to thermodynamics, in particular given the millions of different chemical species and reactions that take place in a cell. (This is distinguished from the large amount of noise in biochemical *measurements* that result from the limitation of experimental techniques like Western Blots for protein expression analysis.) Second, there can be small numbers of certain proteins within cells, making the continuum assumption problematic. This is particularly significant for pacemaker processes because of the importance of transcription factor binding reactions in driving oscillations—and transcription factors are often present only in small numbers within the cell.

Therefore, stochastic, discrete models of chemical reaction mechanisms should be used to investigate the potential for a given biological system model to behave differently if the number of molecules are small and intrinsic noise can have an influence on system dynamics.

Gillespie developed such a stochastic discrete method for chemical reaction modeling that he proved would converge exactly to the deterministic model at the limit of large numbers of molecules [7,8]. The Gillespie method is based on assuming that a reaction between molecules in a reactor is a random process. A model is set up as a set of possible reactions. A reaction is defined by a joint Poisson probability distribution function for both the type of reaction (i.e., molecules from which chemical species react) and the time at which the reaction takes place (which will depend on the reaction rate). Both the reaction type and reaction time will depend on the number of molecules present for each of the chemical species: the likeliest "next" chemical reaction at each time step will be the one that has the fastest rate and the largest numbers of reactant molecules.

It is important to note that the Gillespie method assumes that only one reaction can occur at the same time. As a result, the computational load increases substantially as the number of molecules increases, since reactions will occur close together in time, requiring frequent sampling of the joint probability distribution function. When the Gillespie method started to be popularized for biological system simulation, Gibson and Bruck [6] introduced several straightforward methods to improve the computational performance of the algorithm, which preserve the convergence to the deterministic model at the limit of large molecular numbers. Subsequent work has been done to improve algorithm performance for even larger systems[22]. Other theoretical work has been done to develop hybrid deterministic and stochastic models for cases where there are large numbers of molecules of some chemical species and small numbers for others. However, it has been challenging to prove their convergence to the deterministic continuum limit as well as verify the general accuracy of resulting simulations.

What is particularly significant about stochastic discrete models for biological rhythm modeling is that because of the inherent noise generated within simulations, such models can explore phase trajectories that would be inaccessible to deterministic ODE models. Forger and Peskin[5] applied the Gillespie method to their previous detailed model described above, including the same variables and reaction classes. Among their reported observations was that simulated noise was proportional to $n^{-1/2}$, where n is the number of molecules in the simulation. More biologically significant was a finding that gene duplications reduce the amount of noise in the system and allow it to more robustly generate circadian rhythms. As an example of how models can provide new insights into evolutionary processes, the latter result may account for why there is only one PER locus in the *Drosophila* genome while the mouse genome has evolved to include multiple PER loci.

What was particularly remarkable was that in the stochastic simulation, PER2 mutants were predicted to generate oscillations with the stochastic model and not with the deterministic model. Thus, stochastic and deterministic approaches to modeling circadian rhythm processes can give qualitatively different predictions of system behavior. Stochastic effects may be sufficient to explain unexpected robustness of circadian rhythms found in knockout models. In general, these modeling results exemplify the relative "ease" at which even a simple biochemical model can give rise to oscillations that can drive circadian rhythms, and the robustness of those temporal patterns to noise and external perturbations.

6.8   CONCLUSIONS AND A CAVEAT

Hopefully this brief overview of mathematical modeling of circadian rhythm has introduced you to the terminology used in modeling work, the potential power of modeling, and the limitations of model design and applying model predictions to biological reality. After decades of dormancy while experiments with digital "protein present/absent" experimental results dominated cell and molecular biology, mathematical biology is enjoying a renaissance with the availability of new technologies and an increased focus on the quantitative aspects of complex system behavior. However, it is important to remember that mathematical models, while fun to play with and analyze on their own merits, cannot replace the insight obtained through experimental data. Even more than mathematical physics and chemistry, mathematical biology cannot function independently of experimentation. Thus, we close this chapter with an oft-told joke about mathematical biology:

A mathematical biologist spent his vacation hiking in the Scottish highlands. One day, he encountered a shepherd with a large herd of sheep. One of these cuddly, woolly animals would make a great pet, he thought, so he asked the shepherd, "How much for one of your sheep?"

"They aren't for sale," the shepherd replied. The mathematical biologist pondered this for a moment and then said, "I will give you the precise number of sheep in your herd without counting. If I'm right, don't you think that I deserve one of them as a reward?" The shepherd looked over his vast herd and nodded.

The mathematical biologist confidently said, "387." The shepherd sat silent for a while, sighed, and reluctantly replied: "You're right. I hate to loose any of my sheep. But I'm a man of my word, and as I promised, one of them is yours. Have your pick!" The mathematical biologist grabbed one of the animals, put it on his shoulders, and was about to march on, when the shepherd interrupted: "Wait! Let me guess your profession, and if I'm right, I'll get the animal back."

"That's fair enough," the mathematical biologist agreed. The shepherd immediately declared, "You must be a mathematical biologist." The mathematical biologist was astonished. "You're right," he said. "But how could you know?"

"It's easy," replied the shepherd. "You gave me the precise number of sheep without counting— and then you picked my dog..."

6.9   CHAPTER REVIEW

The chapter began with the recognition that biological rhythms are *emergent properties* which arise from the *complexity of biological systems* through the processes of evolution in an environment dominated by geophysical cycles. This complexity, combined with the sheer number of variables and the difficulties in obtaining accurate measurements of every important parameter, makes modeling an important means of discovering system properties. We began with a *simple mechanical model* that could be solved with *linear differential equations*. That model was able to generate the sustained oscillations characteristics of certain biological rhythms but with an unrealistic caveat—like a frictionless spring, the biological system had to be modeled without any dissipative forces. Nonetheless, this simple model was able to generate some insight into the development of oscillations in systems.

To further approach actual biological systems required moving into the realm of *non-linear systems*. These models, albeit considerably more complex in their mathematical formulations, could also be shown to result in sustained oscillations under certain circumstances. One such circumstance resulted in the development of a *limit cycle*. The models not only established the possibility of such cycles leading to sustained oscillations, but also provided some insight into the nature of those oscillations. One aspect of limit cycle models is their *robustness*—a tendency to return to the original oscillation in the face of slight perturbations.

Next, we investigated selected models of actual biological systems—*activator/inhibitor— feedback loop* approaches—which could be shown to generate limit cycles under certain circumstances. During this discussion, we were able to show how mathematical modeling, such as done by Goodwin[11,12], can anticipate the discovery of the actual molecular mechanisms underlying the biological system being modeled. One aspect of the models that was briefly discussed was the effect of *lag time* within the feedback loops. Different models can generate the same end cycle with different lag times and, in turn, different lag times will generate different frequencies when used within a single model.

Having developed workable models for sustained oscillations, we then turned to the problem of *synchronization*—how can oscillators effect one another in order to generate a common frequency? The approach used here was to concentrate on the problems faced by the so-called circarhythms, biological rhythms which must be adjusted to match the frequency of a geophysical cycle (called a *Zeitgeber*). In this case, only one oscillator can be adjusted—the biological clock—in a process called *entrainment*. The mathematical modeling in this case mirrored the result of the much more rudimentary model of the autorhythmic neuron investigated in Chapter 4 in that entrainment can be achieved through *periodic phase changes*. The resulting map of the effects of the Zeitgeber on the phase of the biological oscillator is called a *phase-response curve* or PRC. Further investigation into the models demonstrated certain limitations on the process of entrainment. Only certain cycles of Zeitgeber and biological oscillator can be synchronized depending on the *strength* (amplitude) *of the Zeitgeber*, *coupling strength* between the Zeitgeber and the biological clock, and the *frequencies of the two cycles*.

As we pointed out, it is not surprising that our mathematical models fit the experimental data. After all, that is what they were designed to do and models which did not fit the data would be rejected as not useful or unrealistic. However, models can also provide unexpected insights, such as when the *waveform* of the entraining signal was examined. Typically, experimental investigations into circadian rhythms use *square waves* as entraining signals. For example, the use of a light/dark cycle of 12 hours of light (provided by various kinds of light sources) following by 12 hours of darkness is very common. This is known as an LD 12/12 cycle. However, when the effects of such a signal were examined, we found that there are unintended consequences. Square waves consist of *multiple harmonics* which can generate conflicting temporal information pushing the entrained biological clocks toward *chaotic behavior*. This does not happen with more gradual sine-wave like signals— such as natural dawn and dusk—and thus there is an *adaptive mismatch* between the experimental signals we investigators often use and the natural geophysical signals to which biological organisms have evolved. This has very practical implications in modern society dominated as it is by artificial lighting.

Continuing with our investigation into entrainment, we found that *phasic entrainment* when combined with *parametric effects* (changes in the fundamental period of biological oscillators) can reposition biological rhythms within an entrained frequency. In other words, the *phase positioning* of rhythms can be altered without losing entrainment. The effect of the T/τ ratio (T for the period of the entraining signal and τ for the period of the entrained signal) is a general phenomenon, applying equally well to geophysical signals entraining biological clocks and to circadian pacemakers entraining secondary oscillators within an organism. Thus, we have discovered another plausible method for both *seasonal changes* and the *organizing of clocks with differing phases* within a single individual. The latter problem was first encountered in Chapter 3 when discussing the problem of *phase angles* between different rhythms in an organism (see Figure 3.10, Chapter 3 for another possible method).

Finally, we began to show how mathematical models could be constructed using modern genomic and proteomics methods to better approximate actual cellular conditions. Again, we were able to show how such models can provide interesting and unintended potential explanations for biological phenomena. For example, when Forger and Peskin[5] applied the Gillespie[7,8] method to their more detailed model, they found that gene duplication reduced the noise and generated a more robust circadian oscillation. This provides a potential explanation for the duplication of the PER locus in mice when compared to the *Drosophila* system.

REFERENCES

[1] Aschoff, A. (1979). Circadian rhythms: Influences of internal and external factors on the period measured in constant conditions. *Zeitschrift fur Tierpsychologie* 49(3): 225–249. DOI: 10.1111/j.1439-0310.1979.tb00290.x 201

[2] Bagheri, N., Lawson, M.J., Stelling, J. and Doyle, F.J. (2008). Modeling the Drosophila melanogaster circadian oscillator via phase optimization. *Journal of Biological Rhythms* 23: 525–537. DOI: 10.1177/0748730408325041 203

[3] Benito, J., Zheng, H., Ng, F.S. and Hardin, P. (2007). Transcriptional feedback loop regulation, function and ontogeny in *Drosophila. Cold Spring Harbor Symposia on Quantitative Biology* 72: 437–444. DOI: 10.1101/sqb.2007.72.009 191

[4] Forger, D.B. and Peskin, C.S. (2003). A detailed predictive model of the mammalian circadian clock. *Proceedings of the National Academy of Science USA* 100: 14806–14811. DOI: 10.1073/pnas.2036281100 202, 203

[5] Forger, D.B. and Peskin, C.S. (2005). Stochastic simulation of the mammalian circadian clock. *Proceedings of the National Academy of Science USA* 102: 321–324. DOI: 10.1073/pnas.0408465102 205, 208

[6] Gibson, M.A. and Bruck, J. (2000). Efficient exact stochastic simulation of chemical systems with many species and many channels. *Journal of Physical Chemistry A* 104: 1876–1889. DOI: 10.1021/jp993732q 204

[7] Gillespie, D.T. (1976). A general method for numerically simulating the stochastic time evolution of coupled chemical reactions. *Journal of Computational Physics* 22: 403–434. DOI: 10.1016/0021-9991(76)90041-3 204, 208

[8] Gillespie, D.T. (1977). Exact stochastic simulation of coupled chemical reactions. *Journal of Physical Chemistry* 81: 2340–2361. DOI: 10.1021/j100540a008 204, 208

[9] Gonze, D. and Goldbeter, A. (2000). Entrainment versus chaos in a model for a circadian oscillator driven by light-dark cycles. *Journal of Statistical Physics* 101: 649–663. DOI: 10.1023/A:1026410121183 193

[10] Gonze, D., Bernard, S., Waltermann, C., Kramer, A. and Herzel, H. (2005). Spontaneous synchronization of coupled circadian oscillators. *Biophysical Journal* 89: 120–129. DOI: 10.1529/biophysj.104.058388 191

[11] Goodwin, B.C. (1963). *Temporal Organization in Cells: A Dynamic Theory of Cellular Control Processes*. Academic Press: New York, London. 187, 191, 206

[12] Goodwin, B.C. (1965) Oscillatory behavior in enzymatic control processes. *Advances in Enzyme Regulation* 3: 425–438. DOI: 10.1016/0065-2571(65)90067-1 187, 191, 206

[13] Hall, J.C. (2003). Genetics and molecular biology of rhythms in Drosophila and other insects. *Advances in Genetics* 48: 1–280. DOI: 10.1016/S0065-2660(03)48000-0 177, 191

[14] Hardin, P. (2005). The circadian timekeeping system of *Drosophila*. *Current Biology* 15(17): R714–R722. DOI: 10.1016/j.cub.2005.08.019 191

[15] Hardin, P. (2006). Essential and expendable features of the circadian timekeeping system. *Current Opinion in Neurobiology* 16(5): 686–692. DOI: 10.1016/j.conb.2006.09.001 191

[16] Konopka, R. and Benzer, S. (1971). Clock mutants of Drosophila melanogaster. *Proceedings of the National Academy of Science* 68: 2112. DOI: 10.1073/pnas.68.9.2112 191

[17] Kronauer, R.E., Czeisler, C.A., Pilato, M., Moore-Ede, M.C. and Weitzman, E.D. (1982). Mathematical model of the human circadian system with two interacting oscillators. *American Journal of Physiology* 242: R3–R17. 184

[18] Lema, M.A., Golombek, D.A. and Echave, J. (2000). Delay model of the circadian pacemaker. *Journal of Theoretical Biology* 204: 565–573. DOI: 10.1006/jtbi.2000.2038 189

[19] Membre, J.-M., Kubaczka, M. and Chene, C. (1999). Combined effects of pH and sugar on growth rate of *Zygosaccharomyces rouxii*, a bakery spoilage yeast. *Applied Environmental Microbiology* 65(11): 4921–4925.

[20] Pittendrigh, C.S. and Daan, S. (1976). A functional analysis of circadian pacemakers in nocturnal rodents. IV. entrainment: pacemaker as clock. *Journal of Comparative Physiology* 106: 291–331. DOI: 10.1007/BF01417859

[21] Reppert, S.M. and Weaver, D.R. (2001). Molecular analysis of mammalian circadian rhythms. *Annual Review of Physiology* 63: 647–676. DOI: 10.1146/annurev.physiol.63.1.647 202, 203

[22] Scheper, T.O., Klinkenberg, D., Pennartz, C. and van Pelt, J. (1999). A mathematical model for the intracellular circadian rhythm generator. *Journal of Neuroscience* 19: 40–47. 189, 205

[23] Slepoy, A., Thompson, A.P. and Plimpton, S.J. (2008). A constant-time kinetic Monte Carlo algorithm for simulation of large biochemical reaction networks. *Journal of Physical Chemistry* 128: 205101. DOI: 10.1063/1.2919546

[24] Sterman J.D. (2008). Risk communication on climate: mental models and mass balance. *Science* 322: 532–533. DOI: 10.1126/science.1162574 187

[25] van der Pol, B. and van der Mark, J. (1928). The heartbeat considered as a relaxation oscillation, and an electrical model of the heart. *Philosophical Magazine, Suppl.* 6: 763–775. DOI: 10.1080/14786441108564652 184

[26] Van Gelder, R.N., Herzog, E., Schwartz, W. and Taghert, P. (2003). Circadian rhythms: In the loop at last. *Science* 300: 1534–1535. DOI: 10.1126/science.1085446 191

[27] Yu, W. and Hardin, P. (2006). Circadian oscillators of *Drosophila* and mammals. *Journal of Cell Science* 119: 4793–4795. DOI: 10.1242/jcs.03174 191

6.10 ADDITIONAL READING

The discussion in this chapter scratches the surface of circadian rhythm models. For more details we point the reader to the following additional (though still limited) selection of journal articles on modeling biological rhythms. Many of the models involve large systems of equations representing the interaction of multiple proteins and genes in molecular networks, and those are included below, along with foundational work and recent reviews. All of the articles listed below are available directly linked from the NIH PubMed database (`http://www.pubmed.gov`), although some may only be available through an institutional subscription.

MORE REFERENCES

[1] Beersma, D.G. (2005). Why and how do we model circadian rhythms? *Journal of Biological Rhythms* 20: 304–313. DOI: 10.1177/0748730405277388

[2] Beersma, D.G., Daan, S. and Hut, R.A. (1999). Accuracy of circadian entrainment under fluctuating light conditions: contributions of phase and period responses. *Journal of Biological Rhythms* 14: 320–329. DOI: 10.1177/074873099129000740

[3] Gammaitoni, L., Hanggi, P., Jung, P. and Marchsoni, F. (1998). Stochastic resonance. *Rev. Mod. Phys.* 70: 223–288. DOI: 10.1103/RevModPhys.70.223

[4] Geier, F., Becker-Weimann, S., Kramer, A. and Herzel, H. (2005). Entrainment in a model of the mammalian circadian oscillator. *J. Biol. Rhythms* 20: 83–93. DOI: 10.1177/0748730404269309

[5] Glass, L. (2001). Synchronization and rhythmic processes in physiology. 410: 277–284. DOI: 10.1038/35065745

[6] Goldbeter, A. (1995). A model for circadian oscillations in the Drosophila period protein (PER). *Proc. R. Soc. Lond. B.* 261: 319–324. DOI: 10.1098/rspb.1995.0153

[7] Goldbeter, A. (2002). Computational approaches to cellular rhythms. *Nature* 420: 238–245. DOI: 10.1038/nature01259

[8] Gonze, D., Halloy, J., Leloup, J.C. and Goldbeter, A. (2003). Stochastic models for circadian rhythms: effect of molecular noise on periodic and chaotic behavior. *C. R. Biologies* 326: 189–203. DOI: 10.1016/S1631-0691(03)00016-7

[9] Granada, A., Hennig, R.M., Ronacher, B., Kramer, A. and Herzel, H. (2009). Phase response curves: elucidating the dynamics of coupled oscillators. *Methods Enzymol* 454: 1–27. DOI: 10.1016/S0076-6879(08)03801-9

[10] Griffith, J.S. (1968). Mathematics of cellular control processes. I. Negative feedback to one gene. *J. Theor. Biol.* 20: 202–208. DOI: 10.1016/0022-5193(68)90189-6

[11] Gunawan, R. and Doyle, F.J. (2007). Phase sensitivity analysis of circadian rhythm entrainment. *J. Biol. Rhythms* 22: 180–194. DOI: 10.1177/0748730407299194

[12] Leise, T.L. and Moin, E.E. (2007). A mathematical model of the Drosophila circadian clock with emphasis on posttranslational mechanisms. *J. Theor. Biol.* 248: 48–63. DOI: 10.1016/j.jtbi.2007.04.013

[13] Leloup, J.C. and Goldbeter. A. (2003). Toward a detailed computational model for the mammalian circadian clock. *Proc. Natl. Acad. Sci. USA* 100: 7051–7056. DOI: 10.1073/pnas.1132112100

[14] Leloup, J.C., Gonze, D. and Goldbeter, A. (1999). Limit cycle models for circadian rhythms based on transcriptional regulation in Drosophila and Neurospora. *J. Biol. Rhythms* 14: 433–448. DOI: 10.1177/074873099129000948

[15] Nakao, M., Yamamoto, K., Honma, K.I., Hashimoto, S., Honma, S., Katayama, N. and Yamamoto, M. (2002). A phase dynamics model of human circadian rhythms. *J. Biol. Rhythms* 17: 476–489. DOI: 10.1177/074873002237141

[16] Ptitsyn, A. (2008). Stochastic resonance reveals "pilot light" expression in mammalian genes. *PLoS ONE* 3: e1842. DOI: 10.1371/journal.pone.0001842

[17] Rand, D.A., Shulgin, B.V., Salazar, D. and Millar, A.J. (2004). Design principles underlying circadian clocks. *J. R. Soc. Interface* 1: 119–130. DOI: 10.1098/rsif.2004.0014

[18] Rust, M.J., Markson, J.S., Lane, W.S., Fisher, D.S. and O'Shea, E.K. (2007). Ordered phosphorylation governs oscillation of a three-protein circadian clock. *Science* 318: 809–812. DOI: 10.1126/science.1148596

[19] Scheper, T.O., Klinkenberg, D., Pennartz, C. and van Pelt, J. (1999). A mathematical model for the intracellular circadian rhythm generator. *J. Neurosci.* 19: 40–47.

[20] Scheper, T.O., Klinkenberg, D., van Pelt, J. and Pennartz, C. (1999). A model of molecular circadian clocks: multiple mechanisms for phase shifting and a requirement for strong nonlinear interactions. *J. Biol. Rhythms* 14: 213–220. DOI: 10.1177/074873099129000623

[21] Smolen, P., Baxter, D.A. and Byrne, J.H. (2001). Modeling circadian oscillations with interlocking positive and negative feedback loops. *J. Neurosci.* 21: 6644–6656.

[22] Ueda, H.R., Hagiwara, M. and Kitano, H. (2001). Robust oscillations within the interlocked feedback model of Drosophila circadian rhythm. *J. Theor. Biol.* 210: 401–406. DOI: 10.1006/jtbi.2000.2226

CHAPTER 7

The Power of Circular Reasoning

Overview

In Chapters 1–6, arguments were presented which support the twin conclusions that: 1. *biological systems are inherently rhythmic* and 2. *the temporal structure of organisms is linked to prominent geophysical cycles.* The data and evidence provided gave ample theoretical and empirical support to these arguments and established that human beings are similar to all other organisms that have evolved within earth's time-varying environment. The *ubiquitous nature of biological cycles*, however, can create difficulties in analysis and understanding. With so many cycles, displaying different frequencies, waveforms, and serving different functions, how are we to make sense of it all?

In this chapter, an attempt will be made to provide an analytical structure—the F^4LM system—by which rhythms can be understood. The acronym stands for *Function, Frequency, waveForm, Flexibility, Level of biological system which is rhythmic, and Mode of rhythmic generation.* The *cardiac cycle* will be briefly used as an example. However, most of the chapter will focus on investigating *neural oscillations* as indicated by the presence of cycles in the *electroencephalogram* or *EEG*. In the process of this investigation, we will discover yet another reason why biological entities *evolved rhythmicity*—the *physical constraints* involved in neural transmission.

As we investigate the nature of neural oscillations, we will come to see that these rhythms are based upon very *specific modifications of the basic membrane systems* described for neurons in Chapter 4. These modifications result in a variety of *autorhythmic neuron oscillators.* The complexity of these modifications and their distribution in neural oscillators throughout the brain argues strongly for *specific evolutionary adaptations* associated with *neural rhythms.* As we continue our investigation, a *model of a limited autorhythmic entities imposing coherent rhythmicity on an entire system* will emerge as a *fundamental pattern* in biological oscillations.

Time discovers truth.

-Seneca, 1st Century Roman Philosopher

Life isn't a matter of milestones but of moments.

-Rose Fitzgerald Kennedy

The mind exists in time, in fact the mind is time; it exists in the past and the future.

-Osho

7.1   BACK TO THE FUTURE

Having just wandered, however briefly, into the world of mathematical models, let us recall what brought us to those models—the ubiquity of biological rhythms. Rhythms and oscillations permeate living systems for a number of reasons, several of which were investigated in Chapters 1–5. To briefly review, we have seen that rhythms are present in living systems because:

1. Feedback control systems are *inherently oscillatory*;

2. Some *activities* that are controlled by biological systems *are themselves rhythmic*, such as walking, breathing, pumping blood, etc.;

3. Complex, goal-oriented *systems require temporal order* in order to function effectively and efficiently;

4. The Earth is dominated by several *prominent geophysical cycles* to which organisms must temporally adapt.

As a result of these factors, biological organisms exhibit multiple levels of rhythmic phenomena operating at various frequencies. The need to maintain both external and internal synchronization, combined with the observation that oscillations cascade throughout complex systems, has led us to conclude that organisms would evolve endogenous or autorhythmic biological clocks to maintain adaptive temporal order. Along the way, we investigated biological rhythms at several different levels, from the action potentials of a single neuron to the cardiac cycle to circadian rhythms and even seasonal cycles. Several common themes emerged, such as the utility of phase response curves to generate entrainment and a model of limited autorhythmic entities imposing coherent rhythmicity on an entire system.

In this chapter, we develop an investigative model for examining biological rhythms, using the cardiac cycle as a prototype example. We then apply this model to the far more complex problem of brain rhythms as typified by the electroencephalogram or EEG. In so doing, we will demonstrate how oscillations can be used to generate temporal order in biological information processing systems.

7.2   A PROTOTYPE INVESTIGATIVE APPROACH

7.2.1   DEFINITIONS AND CLASSIFICATIONS

Consider the following definition for a biological rhythm: *the periodic reoccurrence of an event within a biological system at more or less regular intervals*[15 in 2]. We can approach the classification of these phenomena from several different perspectives. Consider the graph of a generic rhythm from Chapter 2 presented again in Figure 7.1.

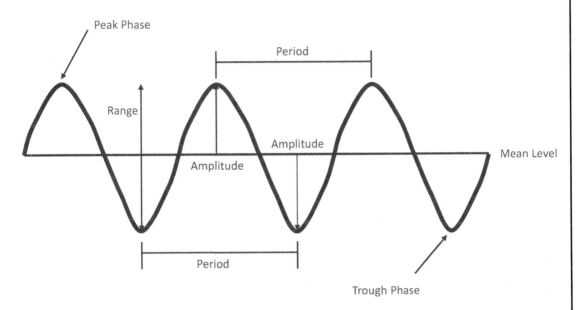

Figure 7.1: Basic graph of rhythm with major parameters provided.

There are several characteristics that we might use to identify or classify a rhythm within the diagram. One is the *period* or *frequency* (the frequency of a rhythm is simply the inverse of the period, e.g., a rhythm with a period of 10 seconds has a frequency of 1 cycle/10 seconds or 0.1 cycles/second). Other characteristics include the *amplitude* above and/or below the mean, the *range*, the mean level around which the rhythm varies, and the peak and trough *phases*.

These are all parameters associated with the rhythm itself and could be considered even if nothing else were known about the phenomenon. There are other characteristics that do not appear in this particular diagram that might also serve to help classify the rhythm, such as *waveform*—the actual patterns of variation—and the *cycle-to-cycle stability*, a measure of how stable the rhythm is in its expression over long periods of time. It is common to use sine or cosine waves to represent biological rhythms. While this is a practice I followed in previous chapters, it can be very misleading if one is not careful. Consider, for instance, the two rhythmic phenomena displayed in Figures 7.2 and 7.3.

These are quite familiar to neuroscientists and cardiologists. Figure 7.2 is a representation of potential changes associated with brain activity as measured by an electroencephalogram (EEG) while Figure 7.3 is a recording of potential changes associated with heart muscle actions as recorded by an electrocardiogram (EKG).

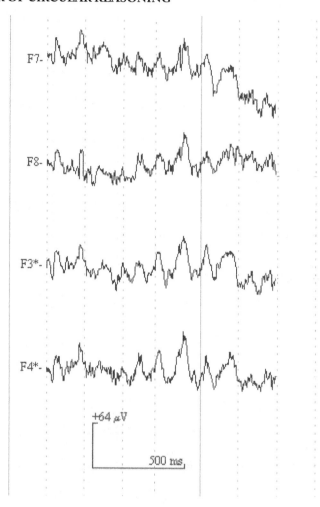

F7-

F8-

F3*-

F4*-

+64 μV

500 ms

Figure 7.2: Scalp electroencephalogram (EEG). (Courtesy of Dr. Scott Bunce.)

Although it would be possible to fit a cosine wave to either of these rhythms, it would require smoothing the various bumps and ripples in the waveforms. These waveforms contain much of the information relevant to understanding the biological cycle we are trying to study. While it is important to measure and understand the fundamental frequency component of the cardiac cycle, analyzing only that aspect of the rhythm eliminates much of the vital data. This loss is illustrated in Figure 7.4 where a cosine wave has been superimposed on the previous EKG recording.

Figure 7.5 displays only the cosine wave and demonstrates the level of data loss in using the cosine representation in place of the actual rhythm. Although critical frequency information remains,

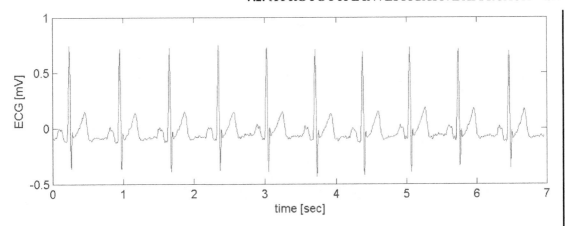

Figure 7.3: Electrocardiogram. (Courtesy of Dr. Chang Chang.)

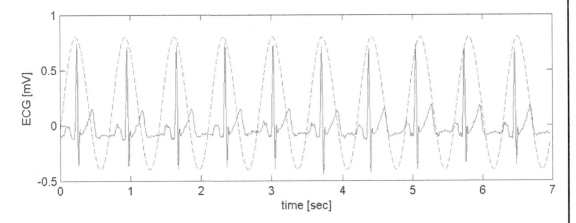

Figure 7.4: Electrocardiogram with fitted cosine. (Courtesy of Dr. Chang Chang.)

much of the most important information lies hidden. For a cardiologist, the *waveform* contains a great deal of the information about a patient's status.

It is unlikely that anyone would suggest reducing the study of EKG patterns to a simple frequency analysis alone. However, this is due to the fact that it is already known what the EKG rhythm represents; in the case of other biological rhythms, the situation may not be so clear. To serve as a guide to the study of such rhythms, I suggest that a classification scheme of some kind is needed, something that can be easily remembered and applied when considering rhythmic phenomena. My approach is to consider each rhythm in terms of four "F"s: *Function*; (wave-)*Form*; *Frequency*; and *Flexibility*. Two other important parameters could be added: the *level of biological system* that is

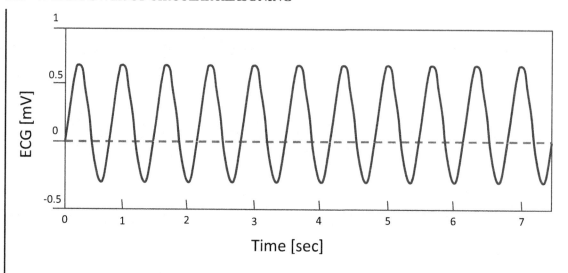

Figure 7.5: Fitted cosine from electrocardiogram without underlying EKG.

rhythmic and the *mode of rhythm generation*. A short mnemonic would then be F⁴LM for function, form, frequency, flexibility, level, and mode. To see how this approach could be applied, consider the EKG pattern.

What is the *function* of the EKG rhythm? Of course, the EKG wave itself does not actually have a function *per se*—it is simply a measure of the underlying electrical activity of the heart during the cardiac cycle. It is this electrical activity that has a function: the sequential muscle contraction which pumps blood throughout the body. If the rhythm ceases, so does the circulation of blood, and the individual dies.

The *form* (or *waveform*) of the EKG indicates various stages of the heart's electrical activity directly tied to the contraction of the cardiac muscle comprising the heart's four chambers. This waveform is fairly complex. There are several components, including the P wave (atrial depolarization), QRS complex (ventricular depolarization and atrial repolarization), and the T wave (ventricular repolarization). The relationship between these components is such that a typical cardiac cycle takes about 0.6–0.8 of a second. This is demonstrated in Figure 7.6.

This brings us to the third F—*frequency*, the inverse of which is called the *period*. One cannot give a single frequency for the heart because the frequency changes depending on the circumstances. Instead, the frequency can be provided as a range from about 60 beats/minute to ~180 beats per minute. The exact range would depend on the individual, although there are some absolute limits— too slow (< 40 beats/minute) and insufficient blood would be pumped through the arteries to deliver adequate oxygen to the brain and keep it functioning; too fast and there would be insufficient time available for the ventricles to fill with blood causing a decrease in stroke volume, the amount of blood

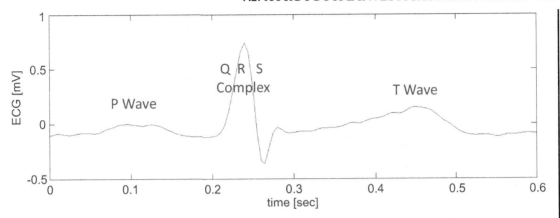

Figure 7.6: Expanded view of electrical potential changes associated with a single cardiac cycle.

pumped out by the ventricles during each cycle. Eventually, at extremely high frequencies, the heart might actually desynchronize, resulting in a loss of coordinated contractions.

This brings us to the fourth F—*flexibility*. The flexibility of the heart seems rather limited and mostly involves changes in frequency. Any large change in the waveform indicates damage or loss of function. For example, a higher amplitude P wave than normal indicates difficulty in moving blood from the atria to the ventricles and might be a sign of an abnormal stiffening of the atrioventricular valves.

What about the *mode of rhythm generation* and *level of the system*? Heart rhythms are generated by specialized cycling cells. These cells are *autorhythmic* in the sense that they generate a cycle on their own without any external cycle driving them. As long as the cells remain alive and are provided with sufficient nutrients, they cycle. Another word that is often used to describe such behavior is *endogenous*. Endogenous rhythmicity can be a characteristic of a cell or tissue as fundamental to that cell or tissue as having a nucleus or being able to contract. In the case of the heart, the autorhythmic cells are bundled into nodes that form a centralized control system. Adaptive changes in the cardiac cycle are managed through this control system to ensure coherent rhythmicity. Thus, the mode of generation of the cardiac cycle—and the EKG which measures it—is *endogenous* at the *tissue level* of organization. When removed from the body, the heart will continue to beat with its own inherent frequency.

The F^4LM classification system provides a kind of snapshot of rhythmic phenomena, a method of organizing what is known about rhythms so that gaps can be recognized or new lines of investigation begun. Hopefully, new insights about the rhythm under study can be generated as well. To see how this might work in a situation where there are more potential gaps to be found, let us turn our attention now to another familiar rhythm, the electroencephalogram or EEG.

7.2.2 THE COMPLEX CASE OF THE ELECTROENCEPHALOGRAM (EEG)

Step 1—What is Being Measured?

In order to truly apply the F⁴LM approach, some understanding of the measurement process itself may be required. In other words, just exactly what is being measured? In the case of the EKG and EEG, there are both biophysical and bio-structural components to be considered. As related to our previous discussion, the biophysical components of the electrocardiogram involve the manner in which the electrical information that is recorded as the EKG is obtained (What potential changes are being recorded? Where are the electrodes placed to record the signal, how is the signal amplified and displayed? And so on.). The bio-structural component is the heart as the living system which generates that information. Although of great interest, an investigation of the biophysical components of either the EKG or EEG is beyond the scope of this text. For such material in relation to EEG, the reader is referred to Nunez and Srinivasan[25]. Still, before proceeding to examine the EEG, we need to determine what type of EEG we are discussing. As it turns out, there are really two categories of EEG—*intracranial EEG* and *scalp EEG*. Intracranial EEG involves implanting electrodes directly into the brain and is far more common in animal studies. Scalp EEG, commonly used in human studies, involves attaching a number of electrodes to various areas of the head.

If we limit our initial discussion to scalp EEGs, the next problem is to understand exactly the bio-structure(s) being measured—what is the underlying rhythm to which a function can be assigned? The answer here is not as straightforward as it was for the EKG. Scalp EEG monitors the summed electrical activity of those neural structures whose potential changes are sufficient to be detected by electrodes placed on the surface of the head. How can one assign a function to that? Is it not just a mess of various signals? While it is true that EEG measurements incorporate a number of signals and are subject to numerous artifacts, there appears to be underlying neural organizations that generate the predominant signals. One of these organizations originates in the communication system linking two processing areas of the brain, the thalamus and the cerebral cortex. To understand the rhythm that results from this system, a simplified introduction to the relationship between the two areas is provided below.

7.2.3 THE SENSORY-THALMACORTICAL SYSTEM

The process by which animals detect environmental information involves the *transduction* of environmental stimuli into action potentials. In engineering terms, a *transducer* is a device—usually electronic—which converts one form of energy into another. All primary sensory neurons can thus be considered as living transducers which convert the energy of an environmental stimulus (light photons, sound pressure waves, chemical scents, etc.) into action potentials. However, since the basic form of all action potentials is identical, how can the brain differentiate between various sensations? Logically, there are only two possibilities. Either the pattern of action potentials or the place to which they are sent within the brain must be linked to specific sensations. The former is called *pattern-coding* and the latter is referred to as *place-coding*. If we think in terms of the Internet, linking to a specific IP address—say your home or Drexel University—is a matter of place-coding while the actual content that is downloaded is a matter of the pattern of transmitted signals, a pattern code. The brain, like the Internet, uses both forms of coding. However, to identify specific sensations, the brain relies on place-coding.

How does this place coding work? With human senses linked to the external environment— with the interesting exception of olfaction or smell—the stimuli are transduced by sensory neurons and this information is sent to an internal brain structure called the *thalamus*. Different areas of the thalamus receive information from different sensory transducers—vision in one area, hearing in another (although there can be processing that occurs in brainstem and midbrain before the thalamus is reached, such as with audition and some body sensations), touch in another, etc. In most descriptions of the system, the thalamus does not seem to engage in any complex processing but rather serves as a combination *relay station* and *gain control device*. With respect to the latter function, the intensity of signals relayed from the thalamus can be adjusted up or down depending on the circumstances. This function is thought to play roles in such phenomena as selective attention and sleep. From the thalamus, signals are forwarded to specific brain areas of the cerebral cortex for more complex processing into the sensory forms we consciously recognize.

This model described above is both one-way (from sense organs to thalamus to neocortex) and passive in the sense that one envisions the neurons waiting passively for input from the environment. In both cases, the model turns out to be inadequate. First, information flow is not one-way. There are certainly thalamic neurons communicating with various neocortical regions (thalamo-cortical neurons) but there are also neurons in the cortex which communicate with the thalamus (corticothalamic neurons). Suppose the interaction was positive in both directions. In such a case, one can easily envision how such a system could generate rhythmic interactions (Figure 7.7). In fact, that is exactly the result one would predict once a positive stimulus interacted with one or the other of the two sets of cells to initiate the process, assuming a delay in processing the information at each end.

The model displayed in Figure 7.7 is extremely simple and uses direct single cell connections to illustrate the potential for rhythmicity. In reality, such connections are through cell networks and involve both positive (excitatory) and negative (inhibitory) links, but the principle remains valid. Each bit of sensory data initiates a response in its own specific thalamocortical loop which then oscillates due to delayed or lagged feedback within the loop. Since different sensations arrive at different times, each loop oscillates at its own phase, different from those of other sensations. The summed measure represented by the scalp EEG over multiple cortical areas would then be the result of many desynchronized or out-of-phase oscillations. Figure 7.8 displays two possible kinds of results from such a model.

The top half (A) of Figure 7.8 shows how these internal oscillations might manifest themselves. The model assumes that each oscillation has the same frequency and differs only in phase. The top curves are the individual oscillations while the bottom shows the summed result of recording all these oscillations simultaneously. Since the EEG is a summed measure, the record might look like the bottom recording if the model is correct. This is a gross simplification since it assumes oscillations

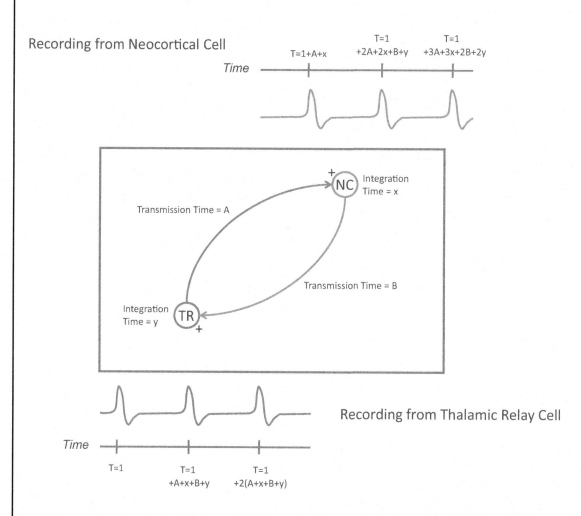

Figure 7.7: Simple model of thalamus-sensory cortex circuit which could result in spontaneous rhythms. In the model presented below, a stimulus is given to the thalamic relay cell (TR) resulting in an action potential at time T=1. After a delay (A) involving the transmission of the signal to the neocortical cell (NC) and another delay in generating an action potential involving synaptic neurotransmission (x), the neocortical cell generates an action potential at time T=1+A+x. This then travels back to the thalamus with transmission delay (B) and after the additional processing/integration delay (y), the thalamic relay cell (TR) generates a new action potential at time T=1+A+x+B+y. Once initiated, such a circuit could maintain a rhythm based upon the connections and delay parameters indefinitely.

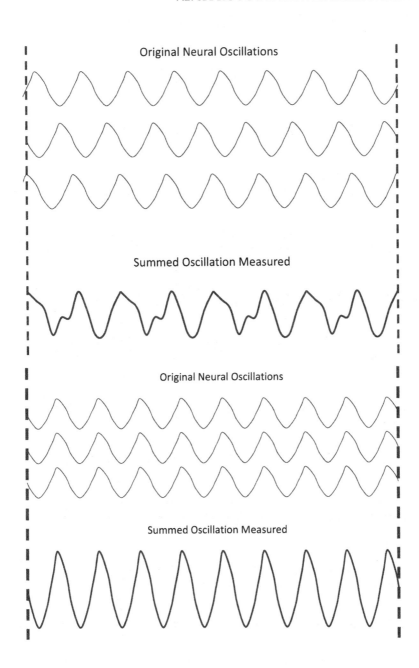

Figure 7.8: Effect of phasing and synchrony on recording scalp EEG rhythms.

differing only in phase but with the same frequency and amplitude. If frequency and amplitude differ as well as phase, the summed result will display an even more complex pattern.

The only change in the situation in going from the top half of Figure 7.8 to the bottom (B) is that the three oscillations were brought into phase alignment with each other. Note that the amplitude has increased and the rhythms are far more clearly defined but the frequency has not been changed a great deal. There is, however, some additional variability in the summed oscillation pattern graphed in A. What emerges from synchronizing the oscillators is a more prominent, coherent, and stable measured rhythm.

Despite the model's simplicity, we can still learn something important from it. Scalp EEG measures the summed potential changes of many neurons. Suppose those neurons are being driven by multiple oscillating systems. The greater the synchrony between those systems, the more coherent the rhythms and the higher the amplitude that will be displayed by the oscillations which appear in the EEG. Thus, some alterations in EEG patterns may be due to increasing and decreasing synchrony between multiple rhythmic systems.

A more realistic problem is diagramed in Figure 7.9 where not only are there phase changes between multiple rhythms but also frequency and amplitude differences as well. How does one go about analyzing these kinds of waveforms? Figure 7.10 shows how such a complex waveform can be decomposed into several simpler waveforms—harmonics—which when added together sum to the original rhythm. This allows the investigator to look for underlying periodicities in complex time series data, such as the EEG, which may indicate various fundamental cycles (see Chapter 6 for discussion). These cycles can then be investigated as separate entities in order to determine their characteristics and function. Some of these analytical techniques by which this can be accomplished will be discussed in Volume 2.

Returning to the EEG, the oscillating patterns and existence of thalamocortical loops demonstrates the inadequacy of applying a unidirectional signaling approach to the processing of sensory information and suggests that EEG patterns are measuring cycling information flow in the system. What about the concept suggesting that the nervous system is a passive receiver of information, waiting for sensory input upon which to act? This perspective has been shown insufficient ever since EEG recordings were first applied to sleeping individuals. Despite low or non-existent levels of sensory input, neural systems continue to be active. EEG patterns change—they do not disappear.

Step 2—Mode and Level of Rhythm Generation

In previous chapters, we have encountered circumstances in which a biological rhythm persists in the absence of any kind of cycling stimulation. The heart discussed above is one obvious example. Taken out of the body, mammalian hearts will continue to beat rhythmically as long as they are kept protected and adequately supplied with energy. Circarhythms are another instance of that phenomenon. Circadian rhythms, for example, have been shown to persist for long periods under constant, non-time-varying environmental conditions. Interestingly, in both these examples, the frequency expressed in isolation is not identical with the frequency expressed under normal cir-

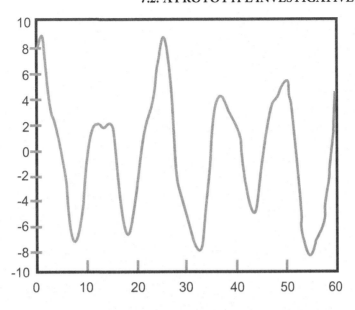

Figure 7.9: Complex waveform from multiple rhythms differing in phase, frequency, and amplitude. (Redrawn from http://www.etsu.edu/math/seier/tperiodogram.doc.)

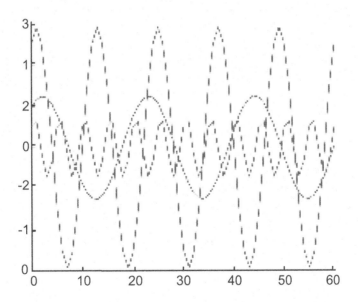

Figure 7.10: Three harmonics which sum to the complex waveform in Figure 7.9. (From http://www.etsu.edu/math/seier/tperiodogram.doc.)

cumstances. In the case of the heart, the frequency expressed outside of the body is typically a bit faster than the cycle expressed when an individual is at rest. For circadian rhythms, the cycle that is normally observed is synchronized to the geophysical cycle of light and dark. When allowed to free-run, the cycle length typically changes. So what frequencies *are* being expressed in isolation? It is the frequencies of the biological pacemakers or clocks which underlie the observed rhythms that are being revealed under these unique circumstances. These are autorhythmic, endogenous living components within each rhythmic biological system that enforce a coherent cycle on the rest of the system. Thus, for both cardiac cycles and circadian rhythms, the *mode of rhythm generation* was based on a limited number of specialized autorhythmic components. The *fundamental level of rhythm expression* was the heart for cardiac cycles and whole organisms for circadian rhythms.

Now if we have evidence that neural oscillations also persist in the absence of external rhythmic stimulation, can we conclude that there must be autorhythmic components to brain systems as well? In other words, what is the *mode* of rhythm generation and at what *level* does it operate? Until recently, the answer has been a bit problematic. Given that scalp EEG measures populations of neurons and because neurons are obligatory oscillators of a sort (see discussion below on how and why neurons became rhythmic), it was always possible that EEGs were the result of a kind of stochastic phenomenon, i.e., random induced oscillations in networks that were not really rhythmic but just appeared to be. The observed EEG patterns could result from the cycling of associated groups of neurons, all of which must oscillate due to the nature of action potentials (again, see discussion below). With millions of induced oscillators firing to process information, waves of potential changes through connected networks are possible, even likely, without any components actually being autorhythmic. To prove this to yourself, recall the effect of a single pulsed stimulation into the fairly simple endocrine feedback system discussed in previous chapters. The result was damped oscillations in the system despite the fact that no component was deliberately made autorhythmic. Figure 7.11 displays this result with the endocrine feedback system.

Now, the actual characteristics of any dampening rhythm depend on the complexity of the network and the type and level of communication between the individual components. Given the enormous complexity and variety of communication in the neural networks of the brain, a longer and more complex dampening cycle is not particularly difficult to envision. Add to this the fact that neural communication is inherently oscillatory (see below) although not necessarily autorhythmic. Assuming that there are multiple neural networks incorporating a variety of communication styles, one could hypothesize that EEG patterns observed in times of reduced environmental stimuli, such as sleep, are the result of dampening rhythms. These rhythms, having varying frequencies and dampening patterns, simply beat in and out of phase with each other, creating the illusion of coherent rhythms. The analysis of harmonics displayed in Figure 7.10 might work with dampening rhythms, assuming the parameters were adjusted properly.

For those of you not adept at adjusting rhythmic parameters in your head, let me give you a simple real-world example. Recall the last time you were in a left turn lane with several cars in front of you. Everyone has their left turn signals on (hopefully!), and you idly watch as your turn signal

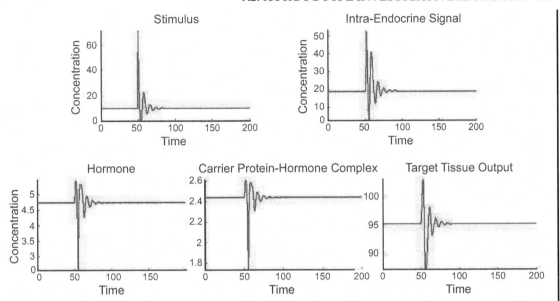

Figure 7.11: System response to a pulse of stimulus applied at Time = 50. Initially the system is at a steady state with no oscillations. However, at Time = 50, a single pulse of stimulus is applied to the system causing a transient perturbation in the system as seen with the damped oscillations in each system component plotted below (stimulus, intra-endocrine signaling molecule, hormone, carrier protein-hormone complex, and target tissue output). The stimulus plotted here is the internal stimulus, X_{in}. (Model and graph by Kevin Freedman.)

beats in and out of phase with that of the car in front of you. This occurs because the two vehicles' turn signals have slightly different frequencies. The turn signals may also have different amplitudes in terms of light intensity as the signals flash on and off. If you were to measure the overall light intensity coming from the combined signals, you would end up with a new rhythm dependent on the varying frequencies and amplitudes of the two underlying signals. This is similar to what we were trying to display in Figure 7.8. One potential hypothesis to explain EEG patterns in the absence of external stimulation is that they are a consequence of summing dampening oscillations in various neural networks. The dampening oscillations are themselves hypothesized to be the result of prior external stimulation and thus, no autorhythmic components are necessary.

The only real way to differentiate between damped and sustained oscillations is to maintain a biological system in a non-stimulatory environment long enough for any damped oscillations to disappear. This is relatively straight forward for organism-level circadian rhythms, and even for many organ and tissue-level rhythms, such as the cardiac cycle. For brain rhythms, this is a different matter entirely. The brain is an exceedingly complex, multiple-purpose system that is not easy to sustain in total isolation from external stimuli. Long-term total sensory deprivation experiments are difficult to arrange and often have deleterious effects, making them ill-suited as investigative tools in this regard[28,39,40]. The alternative is to maintain isolated sections of brain tissue and establish whether or not oscillations can be sustained under those conditions. This requires fairly sophisticated tissue culture techniques and recording apparatus capable of maintaining and recording from brain tissue for sustained periods. As these techniques have become available, they have been used to investigate the question of whether or not autorhythmic components exist in brain tissue. As we will discover a bit later on in this chapter, numerous examples of autorhythmic neurons have indeed been discovered. However, it is important to understand how EEG rhythms might have arisen in the absence of autorhythmic cells to truly comprehend the implications of these discoveries. To accomplish this, a slight digression into the basics of neural transmission is required.

7.2.4   HOW NEURONS BECAME RHYTHMIC

The generation of oscillating signals is not a property restricted to our possible—and at the moment, hypothetical—autorhythmic neurons. Frequency changes in action potentials associated with different levels of stimulation are a property of neurons in general or at least those cells capable of generating action potentials. Pattern coding, as described above, relies in part on this property of frequency changes associated with stimulus intensity. Neurons come in many varieties and vary greatly in their reaction to stimulation. For example, categories of sensory neurons (neurons which transduce environmental stimuli into action potentials) include both tonic and phasic nerve cells. Tonic neurons tend to produce action potentials as long as a stimulus is present while phasic neurons adapt rapidly to stimulation, creating an initial response which disappears quickly even when the stimulus is maintained. Some phasic neurons also show bursts of activity when the stimulus is removed as well, resulting in both "on" and "off" information. This clearly shows that there are other ways of generating oscillatory behavior beyond the environmentally driven negative feedback systems and autorhythmic biological pacemakers or clocks we have already discussed.

However, all of this begs the question of why there is an oscillatory component to neuron function at all, even in tonic neurons whose purpose is presumably to provide information on stimulus intensity and duration. Couldn't this information be transduced and delivered without an oscillatory component? If so, then why do neurons oscillate at all?

Perhaps the first question to ask is why neurons use a frequency coding (action potentials) mechanism to transmit information. Couldn't they have evolved an amplitude-modulated system that would bypass the need to create action potentials? The answer, as almost always is the case with living systems, is yes and no. The reason lies in the nature of currents and the physical reality within whose constraints biological populations must evolve.

Recall the concept of resting and action potentials introduced in a fairly simplified manner in Chapter 4. In review, resting potentials are generated by a very small imbalance of ions across the cell membrane. Action potentials result from ionic currents, mostly in sodium and potassium, crossing the membrane and briefly altering the cellular potential. What was missing from this description was any discussion of how action potentials could be transmitted from place to place. In other words, how does an action potential move along a cell, say from retinal ganglion cell to the thalamus or from the thalamus to the visual cortex (as discussed above)?

To get a signal from place to place, a neuron would have to transmit a change in potential from its point of original to the final destination. One simple method of doing this would be to allow the physical movement of ions generated by an action potential at point A to move to point B. The movement of charged particles, such as ions, is defined as a *current* and is described by Ohm's Law:

$$I = V/R \tag{7.1}$$

where I is the current or movement of charged particles, V is the potential difference driving this movement (measured in volts) and R is the resistance which opposes the movement (measured in ohms).

However, this basic equation does not really help us identify the kind of movements that are possible with ions through cytoplasm. For example, how far would ions travel through a cell for a given change in potential caused by an action potential? To model this, we actually need some concept of the cell structure, since the movement of charged particles through a conducting medium depends on the three-dimensional shape of that medium as well as on the medium's chemical composition. We are fortunate that the structure by which neurons transmit signals is both known and fairly easy to model. The structure is a cylindrical extension of the cell called an *axon* (see Figures 7.12, 7.15, and 7.16).

Classical models of axon dynamics were developed and elaborated by Hodgkin, Huxley, Katz, Miledi, and others. John Nicholls and colleagues have written excellent texts covering these models as part of an overall introduction to functional neuroscience under the title: *From Neuron to Brain*, a 4th edition of which was published in 2001. The equations being used here are from the original text by Stephen Kuffler and John Nicholls[16], published in 1977.

The axon model is based upon three fundamental components: the resistance of neuronal cytoplasm in the axon (or axoplasm), the resistance of the biological membrane defining the axon, and the three-dimensional shape of the axon itself. Depending on the species, the internal specific resistance of axoplasm (R_i) has been reported to range[16,24] from 30 Ω cm to 300 Ω cm; while the specific membrane resistance (r_m) appears to range[16] from 1000 Ω cm^2 to 5000 Ω cm^2, although Meyer, et al.[24] reported a value of 40,000 Ω cm^2. The actual resistance of a cylinder filled with cytoplasm has to be adjusted to the three-dimensional shape:

$$r_i(\text{cylinder}) = R_i L / \pi \rho^2 \tag{7.2}$$

where R is the specific internal resistance of the axoplasm, L is the longitudinal length, and ρ is the radius of the cylinder.

In turn, the distance traveled by the current in relation to a steady state potential change is given by the following equation:

$$V_\chi / V_o = e^{-\chi/\lambda} \tag{7.3}$$

where V_χ is the potential change at a distance χ from the original site where the change occurred, V_o is the potential at the original site and λ is $\sqrt{(r_m/r_i)}$. The value of λ, called the length constant,

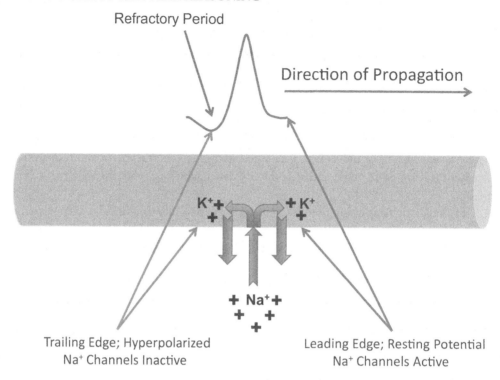

Figure 7.12: Simplified diagram of action potential in an axon.

measures the rate of decline of the potential change and is the distance at which the potential change decreases[16] by $1/e$ or roughly 1/3 of the original value, V_o.

Cross-sectional areas of 1–20 μm are reported for typical mammalian neurons[21]. Suppose we begin with an action potential generating a 100 mV potential change at point A. What would the potential change be at a distance of 1 cm from point A? Plugging in values for R_i of 300 Ω cm, $r_m = 40,000\Omega$ cm^2 and an axon diameter of 20 μm (2×10^{-3} cm), the value of the potential change at a 1 cm distance from A would decrease to a minimal 2.46×10^{-7} mV! The signal generated by a potential change in a living cell attenuates so rapidly as to be an impractical method of long distance communication.

Some of you may be wondering at such a large distance—1 cm—being used on a cellular scale. Animal cells range between 1 and 100 μm in diameter with the typical size being about 10 μm, so, is 1 cm realistic? There are two responses to this objection. First, a quick look at the equations and results indicates that λ will be a fraction of a micron in most mammalian axons, and thus the conclusion that changes in the amplitude of cellular potentials is a poor method of long distance communication remains valid. Second, in many cases axons do have to transmit signals

over considerable distances. Consider communication between the area of the brain involved in generating movements (the motor cortex) and the neurons in the spinal cord that communicate that information to the appropriate skeletal muscles (alpha motor neurons). The axon must stretch from the brain to the spinal cord, which can be many feet in some animals (consider a giraffe or whale communication system). As the calculations above suggest, amplitude changes simply cannot communicate over such distances within the limitations of living systems. That being the case, how is long distance communication achieved?

A simple version of the currents associated with a single action potential in an axon is presented in Figure 7.12. Note that the influx of sodium (Na^+) generates a positive current which then spreads laterally. So far, so good. The problem is that the current does not spread very far and damps out with a few millimeters at best. However, as it turns out, the potential does not have to spread very far. As positive charges accumulate along the adjacent sections of membrane from the site of the action potential, the potential of these sections of membrane rises above threshold. Once above threshold, the feedback system generating an action potential reinitiates in that adjacent section of membrane, and a new action potential is created. This occurs in the next section of axon and again in the next and so on as long as needed until the end of this particular axon is reached. As a result of this rather clever evolutionary innovation, signals can be sent as far as needed without any signal loss at all. Signal transmission might not be blazing fast—1 to 10 meters per second in a non-myelinated axon—but it is extremely reliable.

So we now know why long distance signaling is accomplished through action potentials. Simple potential changes involving ions attenuate much too quickly in normal cytoplasm. It should be noted, however, that sensory transduction is accomplished via amplitude-modulated potential changes. Primary sensory neurons and receptors that react to environmental stimuli do so via graded potential changes—called generator potentials—which are then used to alter the frequency of action potentials. In effect, the nervous system digitizes sensory stimuli by converting graded potential changes into the all-or-none (0's and 1's?) pattern of action potentials. The natural result of this is an oscillatory pattern of on and off in response to stimuli.

Even so, the oscillations of stimulated neurons are not solely the result of action potentials being discrete on/off events. The refractory periods—especially the absolute refractory period when sodium channels are not responsive to any potential changes—dictate a temporal resolution to the frequency of action potentials by enforcing a quiescent period between sequential spikes. Why is this necessary? A quick look at Figure 7.12 provides one explanation. The laterally spreading positive current depicted in the diagram moves in all directions with equal efficiency. In the direction of impulse propagation, the positive current encounters reactive sodium channels. In the opposite direction, due to the refractory periods, the positive current encounters non-reactive sodium channels. In addition, due to the closing of the sodium channels, the outward potassium current through the membrane caused the local potential to be slightly hyperpolarized. Thus, in the forward direction (to the right in this example), the positive current has the opportunity to reinitiate a new action potential, while in the backward direction this possibility is prevented by the refractory periods.

Imagine what would happen if this were not the case. Once an action potential were initiated in an axon, action potentials would be continuously recreated in both directions as long as the conditions permitted. In essence, once an impulse was generated, it would reverberate along the axon until the distribution of ions needed to create an action potential could no longer be maintained. This would hardly be an effective form of communication.

In conclusion, the need for long-distance, discrete, and directional communication dictates a frequency-modulated approach to neural communication. Of course, there are other forms of communication within organisms, such as endocrine signaling through the blood supply, which are not limited by the factors that imposed a rhythmic pattern of action potentials upon neural communication. Even so, endocrine signaling also incorporates a frequency-modulated pattern of communication, the details of which will be examined in Volume 2. In the meantime, you might be wondering if the entire last 3–4 pages were not bit beyond the scope of a discussion on biological rhythms. How does this brief digression into neural signaling relate to our subject?

Recall that one of this text's major themes is that biological rhythms are an inherent characteristic of all living systems. In previous chapters, we discovered that feedback systems are inherently oscillatory, that complex systems require timing that can be accomplished through rhythms, and that adaptation to geophysical cycles generates a cascade of rhythmic responses throughout living systems. Now, we can see that the very physical constraints and opportunities within which life has evolved can dictate a rhythmic response to even the most basic environmental requirements. The inescapable conclusion is that rhythms are an intrinsic part of life and that biological cycles and oscillations are as fundamental as evolution itself to the understanding of living systems.

7.2.5   MODE AND LEVEL OF RHYTHM GENERATION OF BRAIN RHYTHMS REVISITED

We can now see that the very nature of neural transmission imparts an inherently oscillatory character to nerve cells: a frequency-coded action potential is required to transmit directional information over any significant distance using potential changes. We also know from previous chapters that once any component of a complex interconnected system becomes oscillatory, the entire system tends to oscillate in response. So, if neurons are inherently oscillatory and the brain is a highly complex interconnected system, then can we conclude that brain waves, such as EEG patterns, are a side effect of information processing during wakefulness and reflect the dampening complex oscillations during sleep?

As stated above, the final evidence for autorhythmicity in simple neural circuits or even single cells must come from *in vitro* recordings isolated from any driving oscillations. Before proceeding to consider such evidence, let us ask if there is anything more to be learned on this question from scalp EEG.

I am going to argue that there is and that we can use the stages of sleep to predict what might be discovered in the *in vitro* recordings. First, consider the pattern of EEG recordings associated with the various stages of sleep. Typical EEG patterns for these stages are shown in Figure 7.13.

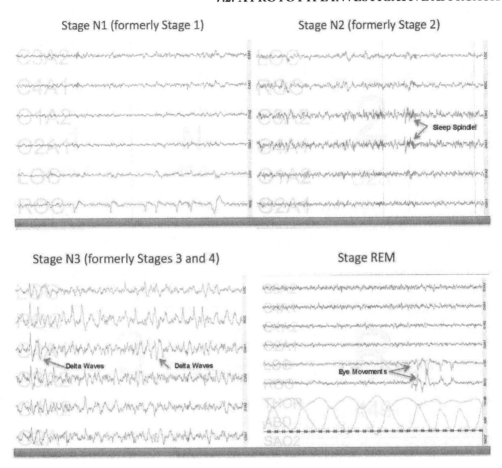

Figure 7.13: EEG patterns of various sleep stages. (Courtesy of Dr. Joanne E. Getsy.)

Sleep can be subdivided into the categories of transitional or light sleep, sometimes called Stages 1 and 2, slow-wave sleep (Stages 3 and 4 or SWS), and rapid-eye movement (REM), also known as active or paradoxical sleep. The EEG pattern of transitional sleep (Stage 1) is dominated by reduced frequency and higher amplitude brain waves of between 3 and 7 Hz (theta) compared with patterns such as alpha (8–12 Hz) and beta (16–25 Hz) typically observed during wakefulness. Stage 2 is characterized by specific patterns: sleep spindle, short bursts of 12–14 Hz activity on a low frequency background, and K complexes, a brief high voltage wave forms with amplitudes often over 100 μV. Deep or slow-wave sleep is characterized by delta waves which are high amplitude, low frequency patterns of 0.5–2.0 Hz[12]. Slow-wave sleep has historically been subdivided into Stage 3 (20% < delta < 50%) and Stage 4 (> 50% delta)[29,30] but recently has been categorized as a single stage[29], N3. REM sleep is called paradoxical for good reason—the EEG pattern shifts toward a mixed frequency of alpha and theta similar to Stage 1 and may even resemble wakefulness. Large areas of the brain are activated, and if it were not for a deliberate blocking of motor impulses at the brainstem level, animals in this state would be acting out the activities experienced in their dreams. All very interesting, but how does this bear on the question of damped vs. sustained rhythmicity?

The frequencies expressed in the brain waves associated with the EEG during sleep range from 25 Hz beta when awake to 0.5 Hz delta when in deep or slow-wave sleep. There is, however, a pattern to how each stage is expressed called the REM/NREM cycle (NREM stands for non-REM and includes all those stages of sleep other than REM). A person when falling asleep passes through stage 1 then 2 then 3 and 4 (slow wave) than back to 2 then into a period of REM. Following this initial REM period, the person returns to stage 3 and 4 again, then back to REM, back down toward SWS, etc.[12] The period of this cycle is highly variable, both between and within subjects,[38] with average adult periods of slightly greater than 100 minutes/cycle. In addition, there is a change in the relative proportions of SWS and REM across the night, with SWS concentrated in the first 1/3 and REM in the final 1/3 of the sleep period. This REM/NREM cycle is displayed in Figure 7.14. Note that this figure uses the older convention of dividing SWS into stages 3 and 4.

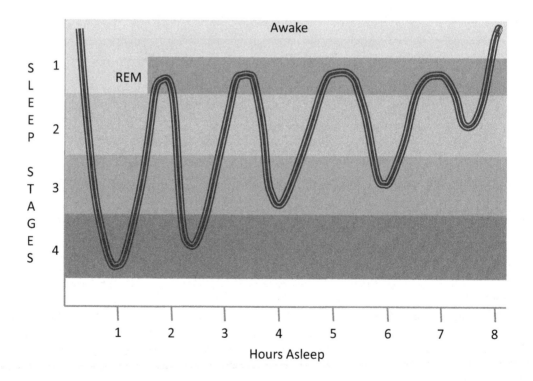

Figure 7.14: Pattern of sleep stages during a typical night's sleep for an adult human.

I have drawn Figure 7.14 in a rather unconventional manner. Instead of the rectangular style normally used[12], I have used a curving line to indicate the REM/NREM cycle. The result displays how one might consider the whole process to be damped oscillation. The diagram *looks* like a damped oscillation, with the initial waveforms encompassing everything from REM to the deepest stages of SWS, while later waveforms are compacted and include mostly REM and light sleep. Although a bit disingenuous—the pattern is highly dependent on the drawing style—it at least shows how it might be possible to imagine how a dampening oscillation in a large scale network is somehow responsible for the brain wave patterns being recorded during sleep.

However, one particular observation makes the damped oscillation hypothesis highly unlikely. In any dampening oscillation, the idea is that the cycles continually decrease in amplitude until the oscillations completely vanish. Figure 7.14 is drawn in such a way as to generate such an impression, although within each stage, the EEG oscillations themselves appear reasonably stable. Since the EEG associated with SWS is of significantly higher amplitude than REM sleep, however, and SWS is seen earlier in the night and REM in the later periods of sleep, the case for dampening oscillations could still be argued. Conversely, it has been reported that individuals who sleep over 12 hours actually show a reemergence of SWS in the later periods of the extended sleeping period[11,37]. True dampening oscillations do not reinitiate themselves and this observation makes it unlikely that EEG patterns during sleep reflect only a dampening of oscillations begun during wakefulness.

7.2.6 THE IN VITRO EVIDENCE

Perhaps the first important issue to be addressed is whether or not neural tissue displays oscillations *in vitro* at all. The answer to that question is yes—a number of studies have conclusively demonstrated that brain slice preparations from various regions and from a number of different species do display sustained oscillations when maintained in tissue culture[1,5,14,17,19,20,27,32,33,34].

The next question concerns the source of these oscillations. Are the cycles the result of some kind of network, as suggested in the simple model shown in Figure 7.7, or are there specific autorhythmic neurons that generate and sustain these rhythms? Generally, the answer to this second question has been that there are autorhythmic neurons involved in generating neural oscillations[1,19,20,27,34]. Interestingly, there are a number of varieties of these specialized cells. For example, there are cells which only display oscillations when experimentally depolarized[27] while others display spontaneous oscillations[20]. Some autorhythmic neurons generate a periodic pattern of action potentials while others display subthreshold oscillations. Some autorhythmic neurons are dependent on fast sodium channels, while others are not[8,17,20,34]. Some oscillating neurons involve calcium currents while others do not[1,18,27]. Thus, not only do endogenously oscillating neurons exist but there is a considerable diversity of mechanisms, styles, and output frequencies among them.

One thing is certain, however—at least some of these neurons show sustained oscillations that do not display any significant dampening. Leresche and colleagues[17] recorded oscillations from individual thalamocortical (TC) neurons using both rat and cat thalamic slices. One category of these autorhythmic cells displayed very large (10–30 mV) spontaneous depolarizations lasted

80–350 milliseconds (generating a delta-wave like oscillation frequency of 0.5–2.9 Hz) with 2–7 action potentials per cycle. Once established, the TC neurons maintained this rhythmicity for as long as the recordings could be sustained, in one case, for 4 hours and 25 minutes. Even if we take the lowest frequency of 0.5 Hz (cycles/second), a sustained rhythm would generate at least 7,950 full cycles in that time period. Using the high end of 2.9 Hz, the result is an astonishing 46,110 repetitions. Moreover, Leresche et al.[17] reported a variation in frequencies of less than 5% during the recording times. Given that a near-24 hour rhythm is accepted as circadian and thus endogenous and autorhythmic if between 25–60 cycles can be observed in temporal isolation, there can be little doubt that dampening rhythms are no longer a viable explanation for neural cycles.

7.2.7   ANOTHER SLIGHT DIGRESSION: IONIC CURRENTS AND NEURAL TRANSMISSION

You may be puzzled by the idea of calcium currents and subthreshold oscillations introduced in the discussion above. Such concepts have not been considered up to now and may seem a bit strange. In order to fully appreciate these autorhythmic neurons, we must now turn our attention to these concepts.

In Chapter 4, the basic concepts of resting and action potentials were developed. As part of the discussion, the ideas of currents, electrical potentials, and conductances were introduced. A membrane crossing current in potassium, I_K, for example, could be calculated using a variation of Equation 7.1 as given below:

$$I_K = g_K(V_M - E_K) \tag{7.4}$$

In fact, this is simply an example of a more general equation that could be applied to any ion capable of crossing the neuron membrane:

$$I_{Ion} = g_{Ion}(V_M - E_{Ion}) \tag{7.5}$$

with the E_{ion} being calculated as was done for potassium and sodium in Chapter 4:

$$E_{Ion} = (RT/zF) * \ln\{[Ion^+]_{out}/[Ion^+]_{in}\} \tag{7.6}$$

The level and direction of current that is generated will obviously depend on the relative concentrations of the ion *in* and *outside* of the cell, the *potential difference* across the membrane and the *conductance*. Given that calcium is typically in higher concentration in the extracellular fluid outside of the neuron compared with the cell interior, and that the resting membrane potential, V_m, is negative inside relative to the cell exterior, one would expect that if calcium were allowed to cross the membrane of a neuron at rest ($g_{Ca^{+2}} \neq 0$), there would be an inward flow of calcium ions. As described in Chapter 4, this would continue until either a balance of forces between diffusion and electrostatics developed or unless some other cellular processes intervened. Under those circumstances, the membrane potential would be dependent not only on sodium and potassium currents, but upon calcium currents as well. In fact, any ion could be modeled in this manner. Whether or not a specific ion actually participates in determining a cell's potential must be investigated empirically, but theoretically, any ion could potentially (pun absolutely intended) contribute.

As it turns out, the discussion of neural potentials in Chapter 4 was exceedingly simplistic. Not only do certain other ions in addition to Na^+ and K^+ contribute to the process—ions such as Ca^{+2} and Cl^-—but also a variety of potassium and sodium channels exist beyond the voltage-gated channels described in Chapter 4. Some of these channels—such as the persistent sodium channel (NaP), the hyperpolarization-activated mixed Na^+/K^+ channel, and the calcium-activated potassium channel—are thought to contribute to neural oscillations[3,6,8,9,35]. However, there are many different channels, and their various roles in brain activity are not fully understood[3,4,36]. Some of the models by which neural oscillations are generated will be briefly discussed below.

In order to understand why subthreshold oscillations matter in neural information processing, we need to understand the means by which neurons communicate with one another. There are two basic mechanisms. The first is a direct connection from one cell to another. Connections between neurons are called synapses, and this particular style of connection is called an *electrotonic synapse*. The connection itself is through a structure known as a gap junction. In a gap junction, protein channels are directly connected from one cell to the other, allowing for the selective flow of ions directly form one cell to the next. This type of connection is displayed in Figure 7.15. The advantage of this method is that it is very rapid and reliable. The disadvantage is that it is somewhat difficult to modify when conditions change.

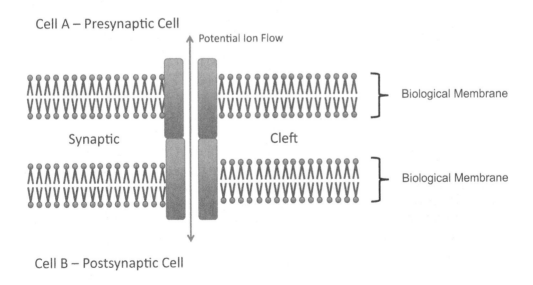

Figure 7.15: Simple model of electronic synapse. In this kind of cell-to-cell connection, chemicals, such as ions, can flow from cell to cell through connected membrane proteins (indicated in light blue).

The alternative type of communication is called a *chemical synapse*, a generalized model of which is presented in Figure 7.16. In this type of communication, a chemical signal is released

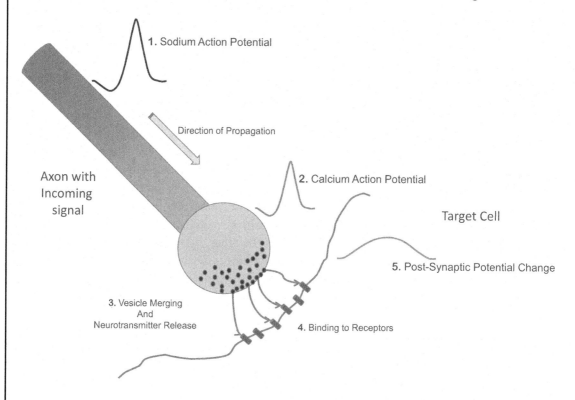

Figure 7.16: Simplified model of chemical neurotransmission. (See text for explanation.)

from the neuron originating a message (being transmitted in the form of action potential) onto a receiving or target cell, often another neuron or muscle cell. The chemical, called a *neurotransmitter*, is stored in the sending (or *pre-synaptic*) cell at the terminal of the axon in small membrane-bound compartments called *vesicles*. When action potential propagates to the axon terminal, it changes from a *sodium-based impulse* to a *calcium-based action potential*. The influx of calcium initiates a cascade of changes within the axon terminal that eventually leads to vesicles moving to the axon membrane and merging with it (a process analogous to two soap bubbles combining). This releases the contents of the vesicles, the neurotransmitters, into the synaptic space (or cleft) to diffuse across the 20–40 nM gap separating the pre-synaptic cell from the target (*post-synaptic*) cell. The neurotransmitter combines with a complex molecule embedded in the post-synaptic cell membrane known as a *receptor*. The neurotransmitter-receptor complex initiates a set of changes in the post-synaptic cells that *transduces* the signal. In many cases, the result of this transduction is a change in the membrane potential of the post-synaptic cell. If the potential becomes more positive, it is called an *excitatory post-synaptic potential* or EPSP (labeled as 5 in Figure 7.16). If the potential becomes more negative, the result is called an *inhibitory post-synaptic potential* or IPSP (not shown).

An EPSP is called excitatory because depolarizing a cell should make it more likely to reach threshold and generate an action potential, while an IPSP is considered inhibitory because hypolarization forces the potential away from threshold, thus making an action potential less likely. This turns out to be a bit simplistic, as will be evident when models for oscillating neurons are discussed, but it will do for now. After transduction, the signal must be *terminated* to prevent excessive or uncontrolled stimulation of the post-synaptic cell. Signal termination (not shown in Figure 7.16) can be accomplished either through metabolic breakdown of the signaling molecule (typical for neurotransmitters such as acetylcholine) or through high affinity re-uptake of the neurotransmitter back into the pre-synaptic axon terminal for reuse (often used for signals such as norepinephrine or serotonin).

7.2.8   RELATIONSHIP TO ELECTROENCEPHALOGRAM (EEG)

The existence of chemical synapses helps to explain the origin of the EEG. In addition to communication lines for outputting information (axons), nerve cells can have multiple cellular extensions specialized for receiving information called *dendrites*. These extensions, along with the cell body itself, are often covered with thousands of chemical synapses, each one of which will generate an ionic current when activated. The sum total of these currents' effects on the cell's membrane potential at the beginning of the cell's axon (called the *initiation segment*) determines whether or not an action potential is generated. In the meantime, these currents, when summed across many neurons, can create changes in the local electrical potential of a specific area. This is recorded as the local mean field potential[10] as diagrammed in Figures 7.18 and 7.19.

Thus, EEG recordings involve the simultaneous and synchronized activity of multiple synapses involving hundreds of neurons over many square centimeters (minimum estimate as 108 neurons in a area of 6 cm^2). For a graphical representation, see Figure 7.19 modified from Olejniczak[26].

7.2.9   MODELS OF OSCILLATING NEURONS

There are many proposed models for the generation of autorhythmic neural oscillations, and the interested reader is referred to Buzsaki[3] and Wang[36] for excellent reviews of the available data. We will introduce some of the concepts associated with the topic by briefly exploring two model systems.

The first model is based upon two additional membrane channels, the hyperpolarization activated mixed cation channel and the persistent, slowly inactivating sodium channel. The currents thus generated are the I_h current and I_{NaP} current. Both are considered to be depolarizing, but have different kinetics and control mechanisms. In the model proposed by Dickson, et al.[8], and Fransen, et al.[9], the primary driver of the oscillation is the I_h current. As the cell hyperpolarizes the I_h channels are activated and a mixed sodium/potassium current (I_h current) begins to flow. Since the channel allows both major monovalent cations to cross, the reversal or equilibrium potential is

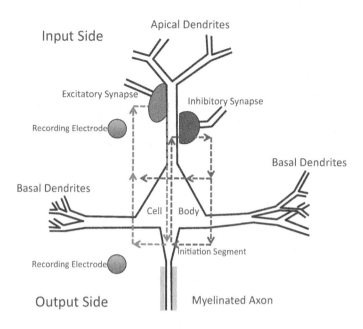

Figure 7.17: Diagram of cortical pyramidal cell showing the effects of excitatory and inhibitory synapses on current flows across the cell membrane. (Redrawn from Freeman[10].)

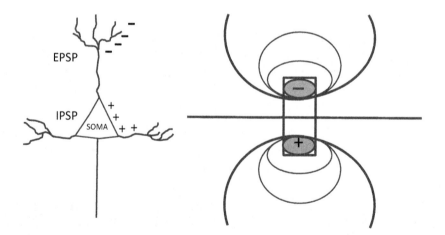

Figure 7.18: Diagram of pyramidal cell (left) with an electrical dipole representation of the effects of current flows. (Redrawn from Tatum, et al.[31].)

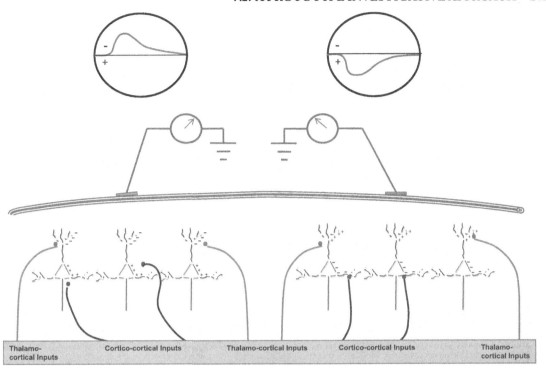

Figure 7.19: Generation of potential fields from synaptic activity for recording by scalp EEG. (Redrawn from Olejniczak[26].)

between the equilibrium potentials for sodium (E_{Na}) and potassium (E_K). From the discussion in Chapter 4, we know that E_{Na} is highly positive (usually between +50 and +60 mV) while E_K is slightly negative relative to the resting potential. Thus, the combined current moves the cell's potential in a positive direction. However, as the cell begins to depolarize, the h channels close, the I_h current stops and the cell begins to hyperpolarize. As the cell hyperpolarizes, the h channels start to reopen and the process begins again. Observations of the effects of h channels on cell potentials demonstrate that the h channels contribute to the resting potential by providing a slight sodium current. Bear in mind from Chapter 4 that the resting potential is slightly more positive than E_K due to such a slight sodium current. Apparently, the h channels in some oscillating neurons contribute to resting potential such that a complete blockade generates hyperpolarization (recall that when all sodium channels are blocked during a typical action potential, the remaining potassium current briefly hyperpolarizes the cell). Thus, all that is necessary to generate a cycling in the membrane potential is a hyperpolarized activation of h channels, followed by a subsequent depolarization which then inactivates them, resulting in hyperpolarization and so on.

In that case, what is the purpose of the NaP channel and current? Apparently, these channels serve as a lag control mechanism that modulates the active-inactive h channel cycle to generate specific frequency ranges. Note that the activation of NaP (g_{NaP}) channels is slightly out-of-phase with the h channel activation (g_h). The result of this is to generate greater levels of depolarization and hyperpolarization, and thus greater changes in the h channel states, than would be possible with just the h channel activation-inactivation cycle[8,9]. The resulting cycling conductances and potentials are diagrammed in Figure 7.20.

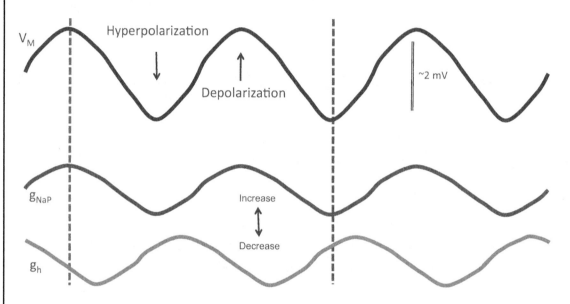

Figure 7.20: The relationship between the conductances of the hyperpolarized mixed cation channel (g_h) and the persistent sodium channel (g_{NaP}) on the generation of subthreshold oscillations in membrane potential. (Redrawn from Dickson, et al.[8]; Fransen, et al.[9].)

All of this begs a rather important question: *What is the purpose of subthreshold oscillations in the first place?* It makes some sense to generate oscillations in neurons if those cycles lead to actual signals, i.e., action potentials. However, if there is no signal being generated, what is the point? Is this just some sort of strange evolutionary artifact of neuron design or is there an adaptive function to this cycling?

To understand how oscillations in the subthreshold range affect neuron information processing, first recall that such complex processing requires multiple neurons and thus neuron-to-neuron communication. For reasons discussed above, not all such communication can go through electronic channels. For higher levels of flexibility and adaptability, chemical synapses are needed. Now, as displayed in Figure 7.16, neurotransmission through a chemical synapse often alters the membrane potential of a target cell. These effects can lead to either depolarization (excitatory post-synaptic potentials abbreviated as EPSP) or hyperpolarization (inhibitory post-synaptic potential abbreviated as IPSP). For an EPSP to generate an action potential in a target neuron, for example, it must be large enough to exceed the threshold. At this point, things can get very complicated. A single EPSP is seldom enough to generate an action potential in a target neuron. So, there are two main alternatives to get that result. A single sending neuron could generate a series of action potentials (APs) so closely aligned in time that their EPSPs begin to merge on top on one another. In other words, before the target can return its membrane potential to the resting potential, another EPSP is generated, depolarizing the membrane still further. If enough EPSP are generated sufficiently close together in time, the *temporal summation* of EPSPs can exceed threshold and an AP is produced in the target. This model envisions one sending (pre-synaptic) cell firing repeatedly. Another method would be to have several input or pre-synaptic cells connected to a single target or post-synaptic cell. Then, if all the input cells fired simultaneously (in synchrony?), the number of EPSPs produced might exceed threshold and the target cell fire an AP. This model is called *spatial summation*. Given that the estimated number of neurons in the human brain is somewhere on the order of 10^{15} and each of these cells can have between 10,000 and 100,000 synaptic connections, the sheer raw processing power of the structure is awe-inspiring.

As stated previously, complex devices require considerable temporal coordination, and the brain is no exception. Enter the subthreshold oscillations of autorhythmic neurons. Consider the two possibilities displayed in Figure 7.21. In the top graph, the situation with a passive neuron is displayed. This has been the traditional way to describe the effects of neurotransmission for many years. The bottom graph shows the same stimulation applied to an autorhythmic neuron displaying subthreshold oscillations. The results are very different—in fact, despite being exposed to the same input stimuli, only one output—the last AP—is the same in the two circumstances. Note that the subthreshold oscillations can either increase or decrease the likelihood of generating an AP in the target cell. As a result, the action potentials generated in the target oscillator are produced with a characteristic frequency and phase. Thus, the autorhythmic neuron acts as a kind of temporal filter in both frequency and phase, translating a potentially random set of inputs into a temporally coordinated output.

The second model uses a bit of the first (the h channel and I_h current) but adds a new wrinkle in the form of a potential-sensitive calcium channel, the T channel. In this new model, the process begins when the cell is hyperpolarized below a certain value, which is -65 mV in Figure 7.22. As in the first model, hyperpolarization opens the mixed cation h channel, leading to a slow depolarization. However, there is a new potential-sensitive element, the T channel, which opens at a specific potential range. Thus, as the mixed sodium/potassium current (I_h) raises the membrane potential, the h channels are inactivated (closed) while the T channels are activated (opened). Since the equilibrium or reversal potential for calcium is very positive under these circumstances (considerably higher than that for sodium), depolarization continues even with the h channels now shut. As calcium flows into the cell (I_T current), the T channels begin to close and a calcium-activated potassium channel opens. It is interesting that there are so many ways to control membrane conductances and currents

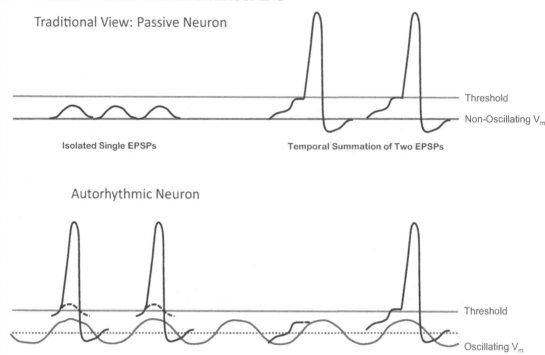

Figure 7.21: Effects of neurotransmission on passive neuron (top) and autorhythmic neuron with subthreshold oscillations.

through channel proteins—potential sensitive channels, chemical signal-sensitive channels, and now ion-sensitive channels. Activation (opening) of the calcium-sensitive potassium channel increases potassium conductance (g_K), leading to an increase in an outward-directed potassium current (I_K) which, in turn, hyperpolarizes the cell, thus reinitiating the cycle. The major steps in this oscillation are linked together with red arrows in Figure 7.22.

The nature of these two models is significantly different in terms of output. The subthreshold oscillator operates more as a kind of temporal filter, a method of changing temporally random inputs into outputs with specific frequencies and phases. This allows for very specific temporal coordination of information processing. However, without a continuous supply of input messages, other neurons in the subthreshold oscillator's network would be unaware of the oscillator's cycle. Once initiated, a subthreshold oscillator could continuously cycle in the background of a quiescent network without any tell-tale signs that a cycle was running at all.

That is not the case with the second model. The influx of calcium depolarizes the neuron above threshold and initiates a series of sodium-dependent action potentials. In effect, the calcium current serves as a depolarizing input as was discussed in Chapter 4. As you may recall, stimulating

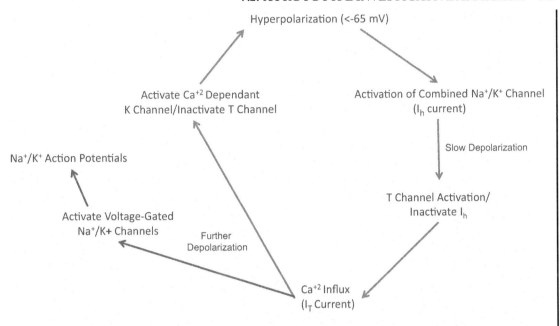

Figure 7.22: Model for oscillatory neuron. The main steps in generating and maintaining the oscillation are linked via red arrows. The output from the oscillation is indicated by the blue arrows. (Based on material from Destexhe, et al.[6].)

a neuron with a constant depolarizing stimulus above threshold leads to a sustained burst of action potentials at a frequency dependent on both the level of depolarization generated and the specific characteristics of the neuron, such as the exact length of the refractory periods. In this model, the calcium current depolarizes the neuron above threshold for a sustained period, leading to a burst of sodium-dependent action potentials. Thus, this model actually incorporates two different cycles with different frequencies. There is the main oscillation generated via the h and T channels and I_h and I_T currents (red arrows in Figure 7.22) and a second rapid oscillation generated by the action potentials when the main oscillation goes through its period of sustained depolarization (blue arrows in Figure 7.22).

Although these kinds of models are significantly more complex, with many more variables and parameters than the basic approaches discussed in Chapter 4 and above, they are based upon the same fundamental principles. For example, two of the new equations used to generate the second model described above are[6]:

$$I_h = g_h S_1 F_1 (V_m - E_h) \tag{7.7}$$
$$I_T = -g_{Ca} m^3 h (V_m - E_{Ca}) \tag{7.8}$$

Despite the fact that here are several new parameters to be considered in terms of the speed and sensitivities of activating and inactivating these new channels (S_1, F_1, m and h), the basic form is the same as Equation (7.5) which was, in turn, derived from Ohm's Law. The new channels and currents have not been magically created from nothing—fundamentally, they are modifications of the basic approaches from which neurons originally evolved. We see here another example of the explanatory power of the basic evolutionary principle of descent with modification.

We have now answered the question of the *mode* by which many neural rhythms are produced. The basic rhythm-generating system in these cases appears to be autorhythmic neurons. Such neurons are widely distributed throughout mammalian brains having been found in the amygdala[27], entorhinal cortex[8], neocortex[5], thalamus[7,17], ventral tegmentum, and substantia nigra pars compacta[23], midbrain dopaminergic cells[13], and inferior olive[19]. In fact, rhythmic neurons have been found in the peripheral nervous system as well, such as in amacrine cells of the retina[34]. This does not, however, get us to the *level(s)* of rhythm expression. After all, we began this discussion considering the patterns obtained with scalp EEG, and given that scalp EEG measures population phenomena and not single neurons, the question remains as to how one gets from one to the other. Actually, this question has already been answered in other forms in previous chapters. Recall that when numerous elements are linked together in feedback systems, when one element becomes rhythmic, the entire system tends to oscillate in response. We have seen this with endocrine feedback models and described it as the oscillatory cascade that runs from the largest macro-scale geophysical cycles to the smallest micro- and nano-scale biochemical cycles. Link multiple components together in an interacting network, make one component oscillate, and get system-level rhythms. It seems this principle is again at work in neural networks, where autorhythmic neurons help to drive network oscillations. This was ably demonstrated in a model of thalamic network oscillations where the network remained stable and quiescent until one element was made to oscillate at which time the entire network became oscillatory[7]. The basic model that emerges from these observations is that of a distributed network of local autorhythmic neurons driving oscillating neural assemblies. These assemblies interact and the effects of their interactions on cortical neurons is what is being measured by the scalp EEG.

7.2.10   DISTANCE AND LAG TIME

There is an additional piece of this rhythmic puzzle yet to be discussed which concerns *distance* and *lag time*. The question arises from the problem of signal transmission time and the potential impact such times might have on the maintenance of synchrony in neural assemblies. We can conveniently consider the issue in terms of the distribution of synchronized neural assemblies within a brain as encompassing *local* and *global* networks.

Neural assemblies located within small regions of the brain can be thought of as the equivalent of a *Local Area Network* or LAN. Such LANs in the brain may have minimal lag times associated with neural transmission due to short distances between cells and may involve limited numbers of chemical synapses as well. In fact, neurons may have even electrotonic synapses to generate coordinated rhythm activity within very small brain regions[14]. This is what the heart does with its two linked LANs, the atrial and ventricular systems. However, the heart has very limited information processing capabilities, and electrotonic synapses are not well suited for the modifications associated with complex, high-level processing (learning, memory, etc.). Thus, one should normally expect significant numbers of chemical synapses in neural tissue, even within LANs.

This is probably not a significant issue for localized networks but could be a problem for maintaining rhythms across larger, more global distances (i.e., *Wide Area Networks* or WANs). There are really two issues. First, there is the lag generated by the act of impulse propagation itself. The longer the axon, the more time it takes to get from the initiation segment to the axon terminal. This may not seem to be too great a problem. Propagation speeds can range from 1 to 100 meters per second in typical mammalian axons depending on size and myelination. However, with oscillations in the 40–100 Hz range, even these propagation speeds could generate slight delays which might impact the ability of a dispersed neural network to maintain synchrony. Even so, a constant and consistent volley of APs could generate network synchrony when maintained for a sufficiently long period if it were not for a second issue: the *lags* and *noise* generated by chemical neurotransmission.

Relative to impulse propagation, chemical neurotransmission is a complex process involving a number of biochemical and biophysical events, including calcium influx, calcium binding, and a cascade of reactions leading to vesicle fusion, diffusion of the signal across the synaptic cleft between sending and receiving cell, binding of signal to receptor, transduction by the signal-receptor complex into a biological effect, etc. (see Figure 7.16) This creates both lag and noise. The *lag* comes from having numerous chemical processes, each of which requires a certain amount of time, all of which must be accomplished in order to obtain the end result on the target cell. The *noise* also comes from the involvement of many processes, each of which has a certain stochastic character determined by its molecular interactions, how the molecules involved diffuse, etc. We can think of each of these noisy processes contributing a bit of "jitter" into the overall process[22]. Although no process alone is going to generate large amounts of such jitter, the combination of processes, each of which adds a small but definitive level of random variation, may result in a finite level of noise being introduced into the transmission of a signal from one cell to another.

Thus, the problem is not really the physical distance the information must travel but the number of chemical synapses involved in the network. The larger and more dispersed the network, the greater the number of chemical synapses involved and the greater the probability of desynchronization through a combination of lags and noise. It is possible that controlling for such issues is the reason that autorhythmic cells are found distributed throughout the brain. After all, without noise, even a large-scale network should be able to be forced into coherent oscillations by a very few, or even one, component. Why then are there so many autorhythmic neurons? One possibility is that these cells act as both *temporal repeaters* and *filters* on the network. Even subthreshold oscillations can both filter out noise and reinforce a specific temporal sequence (see Figure 7.21). By having distributed autorhythmic neurons in various networks, specific temporal patterns and rhythms can be maintained in the inherently noisy world of chemical neurotransmission. It is also possible, even likely, that different oscillators and cycling networks serve different processing functions in the brain. The need for complex information processing could easily require multiple networks operating at different frequencies.

Pay close attention to these last statements. Neural oscillators are clearly more complex that the very simple model of the neuron originally presented in Chapter 4. There are a number of additional membrane channels with characteristics that make those channels uniquely suitable for the generation of specific cycles. In fact, the variety of different neural oscillators and the various control systems which have evolved to generate different frequencies and outputs do not support the hypothesis that such oscillations are merely a stochastic phenomenon arising from the firing of large numbers of relaxation oscillators (typical neurons). Rather, the patterns reveal that oscillation itself must play a fundamental role in the information processing of the nervous system.

7.3   CHAPTER REVIEW

We began this chapter with an overview of biological rhythms and a suggestion of how such rhythms might be characterized—the F^4LM approach (function, form, frequency, flexibility, level, and mode). After a brief discussion of cardiac rhythms, we began to investigate the complex issues underlying the EEG. In the process of that investigation, we made a number of observations. These included:

1. The mechanisms of action potential generation and propagation force neurons to be relaxation oscillators;

2. A number of neurons have evolved into autorhythmic cells, generating a multiplicity of oscillations under various conditions using a diversity of mechanisms;

3. These autorhythmic neurons can drive network level oscillations.

As a result, we now have partial answers to the questions of mode and level of rhythm generation in the brain. As with many biological rhythm systems, neural rhythms appears to depend on a limited number of autorhythmic components driving a broad network of associated elements. In this case, a widely distributed set of specifically evolved autorhythmic neurons generates oscillations within connected neural assemblies. The exact nature of these assemblies remains to be determined, as do the other parameters of the F^4LM approach—function(s), form(s), frequency(ies), and flexibility. These will be discussed in the subsequent volume of the series, Volume 2.

Let us review those major factors promoting the evolution of biological rhythmicity as previously discussed and demonstrated. These factors are:

1. Feedback control systems are *inherently oscillatory*;

2. Some *activities* that are controlled by biological systems *are themselves rhythmic*, such as walking, breathing, pumping blood, etc.;

3. Complex, goal-oriented *systems require temporal order* in order to function effectively and efficiently;

4. The Earth is dominated by several *prominent geophysical cycles* to which organisms must temporally adapt.

During our investigation of neuron function, we were able to add yet another reason, physical constraints. This leads to the fifth factor:

5. Physical and chemical parameters may so *constrain a biological system* that *oscillations become an inevitable consequence.*

In examining these factors, it would seem most likely that neural oscillations are due to the evolutionary impact of Factor 3—that complex systems require temporal order. The other factors do not seem to offer as good an explanation for brain rhythmicity. Information processing *per se,* such as accomplished by the brain, does not seem to be rhythmic in nature—one could easily imagine information systems which do not oscillate, reducing the potential impact of Factor 2. Neural networks do involve feedback mechanisms but do not seem to have evolved primarily to maintain a certain set point, as was the case in the model of the endocrine system described previously. Information processing is far more complex an enterprise and this decreases the potential explanatory power of Factor 1. There appear no obvious geophysical cycles with frequencies in the 0.5–200 Hz range to which neural oscillators are entraining, so Factor 4 does not appear to be a satisfactory explanation. One should keep in mind, however, that the lack of known geophysical cycles in these frequency ranges does not preclude the discovery of such cycles at some future time. Finally, despite the fact that we uncovered Factor 5 specifically associated with neural function, the existence of very precise and detailed mechanisms for creating oscillations within neurons above and beyond anything dictated by physical or chemical constraints strongly argues for an evolutionary adaptation and selective advantage for neural oscillations. Thus, as we turn our attention to the function of neural oscillations in Volume 2, we anticipate these functions to involve the need for temporal order in complex systems.

REFERENCES

[1] Amitai, Y. (1994). Membrane potential oscillations underlying firing patterns in neocortical neurons. *Neuroscience* 63(1): 151–161. DOI: 10.1016/0306-4522(94)90013-2 235

[2] Aschoff, J. (1981). A survey on biological rhythms. In Aschoff, J. (ed.) *Handbook of Behavioral Neurobiology 4: Biological Rhythms.* Plenum Press: New York, pp. 3–10.

[3] Buzsaki, G. (2006). *Rhythms of the Brain.* Oxford University Press: New York. DOI: 10.1093/acprof:oso/9780195301069.001.0001 237, 239

[4] Buzsaki, G. and Draguhn, A. (2004). Neural oscillations in cortical networks. *Science* 304: 1926–1929. DOI: 10.1126/science.1099745 237

[5] Compte, A., Reig, R., Descalzo, V.F., Harvey, M.A., Puccini, G.D. and Sanchez-Vives, M.V. (2008). Spontaneous high-frequency (10–80 Hz) oscillations during up states in cerebral cortex

in vitro. *The Journal of Neuroscience* 28(51): 13828–13844.
DOI: 10.1523/JNEUROSCI.2684-08.2008 235, 246

[6] Destexhe, A., Babloyantz, A. and Sejnowski, T.J. (1993). Ionic mechanisms for intrinsic slow oscillations in thalamic relay neurons. *Biophysical Journal* 65: 1538–1552.
DOI: 10.1016/S0006-3495(93)81190-1 237, 245

[7] Destexhe, A., Bal, T., McCormick, D.A. and Sejnowski, T.J. (1996). Ionic mechanisms underlying synchronized oscillations and propagating waves in a model of ferret thalamic slices. *Journal of Neurophysiology* 76(3): 2049–2070. 246

[8] Dickson, C.T., Magistretti, J., Shalinsky, M.H., Fransen, E., Hasselmo, M.E. and Alonso, A. (2000). Properties and role of I_h in the pacing of subthreshold oscillations in entorhinal cortex layer II neurons. *Journal of Neurophysiology* 83: 2562–2579. 235, 237, 239, 242, 246

[9] Fransen, E., Alonso, A.A., Dickson, C.T., Magistretti, J. and Hasselmo, M.E. (2004). Ionic mechanisms in the generation of subthreshold oscillations and action potential clustering in entorhinal layer II stellate neurons. *Hippocampus* 14: 368–384. DOI: 10.1002/hipo.10198 237, 239, 242

[10] Freeman, W.J. (1992). Tutorial on neurobiology: From single neurons to brain chaos. *International Journal of Bifurcation and Chaos* 2(3): 451–482. DOI: 10.1142/S0218127492000653 239, 240

[11] Gagnon, P. and De Koninck, J. (1984). Reappearance of EEG slow waves in extended sleep. *Electroencephalography and Clinical Neurophysiology* 58: 155–157.
DOI: 10.1016/0013-4694(84)90028-2 235

[12] Hobson, J.A. (1989). *Sleep*. Scientific American Library: New York. 233, 234, 235

[13] Hyland, B., Reynolds, J.N.J., Hay, J., Perk, C. and Miller, R. (2002). Firing modes of midbrain dopamine cells in the freely moving rat. *Neuroscience* 114(2): 475–492.
DOI: 10.1016/S0306-4522(02)00267-1 246

[14] Jacobson, G., Rokni, D. and Yarom, Y. (2008). A model of the olivo-cerebellar system as a temporal pattern generator. *Trends in Neuroscience* 31(12): 617–625.
DOI: 10.1016/j.tins.2008.09.005 235, 247

[15] Kalmus, H. (1935). Periodizitat und Autochronie (= Ideochronie) als zeitregelnde Eigenschaffen des Organismus. Biologia Generalis 11: 93–114.

[16] Kuffler, S.W and Nicholls, J. (1977). *From Neuron to Brain*. Sinauer Associates: Sunderland, MA. 229, 230

[17] Leresche, N., Lightowler, S., Soltesz, I., Jassik-Gerschenfeld, D. and Crunelli, V. (1991). Low-frequency oscillatory activities intrinsic to rat and cat thalamocortical cells. *Journal of Physiology* 41: 155–174. 235, 236, 246

[18] Llinas, R. (1988). The intrinsic electrophysiological properties of mammalian neurons: Insights into central nervous system function. *Science* 242: 1654–1664. DOI: 10.1126/science.3059497 235

[19] Llinas, R. and Yarom, Y. (1986). Oscillatory properties of guinea-pig inferior olivary neurones and their pharmacological modification: An *in vitro* study. *Journal of Physiology* 376: 163–182. 235, 246

[20] Llinas, R.R., Grace, A.A. and Yarom, Y. (1991). *In vitro* neurons in mammalian cortical layer 4 exhibit intrinsic oscillatory activity in the 10- to 50-Hz frequency range. *Proceedings of the National Academy of Science* 88: 897–901. DOI: 10.1073/pnas.88.8.3510c 235

[21] Malmivuo, J. and Plonsey, R. (1995). *Bioelectromagnetism*. Oxford University Press: New York. 230

[22] McAuley, J.H. and Marsden, C.D. (2000). Physiological and pathological tremors and rhythmic central motor control. *Brain* 123: 1545–1567. DOI: 10.1093/brain/123.8.1545 247

[23] Mecuri, N., Bonci, A., Calabresi, P., Stratta, F., Stefani, A. and Bernardi, G. (1994). Effects of dihydropyridine calcium antagonists on rat midbrain dopaminergic neurons. *British Journal of Pharmacology* 113: 831–838. DOI: 10.1111/j.1476-5381.1994.tb17068.x 246

[24] Meyer, E., Muller, C.O. and Fromherz, P. (1997). Cable properties of dendrites in hippocampal neurons of the rat mapped by a voltage-sensitive dye. European Journal of Neuroscience 9(4): 778–785. DOI: 10.1111/j.1460-9568.1997.tb01426.x 229

[25] Nunez, P.L. and Srinivasan, R. (2006). Electric Fields of the Brain: *The Neurophysics of EEG*. Oxford University Press: New York. DOI: 10.1093/acprof:oso/9780195050387.001.0001 220

[26] Olejniczak, P. (2006). Neurophysiologic basis of EEG. *Journal of Clinical Neurophysiology* 23(3): 186–189. DOI: 10.1097/01.wnp.0000220079.61973.6c 239, 241

[27] Pape, H.-C., Pare, D. and Driesang, R.B. (1998). Two types of intrinsic oscillations in neurons of the lateral and basolateral nuclei of the amygdala. *Journal of Neurophysiology* 79: 205–216. 235, 246

[28] Schuklman, C.A., Richlin, M. and Weinstein, S. (1967). Hallucinations and disturbances of affect, cognition, and physical state as a function of sensory deprivation. *Perceptual and Motor Skills* 25(3): 1001–1024. DOI: 10.2466/pms.1967.25.3.1001 228

[29] Silber, M., Ancoli-Israel, S., Bonnet, M., Chokroverty, S., Grigg-Damberger, M., Hirshkowitz, M., Kapen, S., Keenan, S., Kryger, M., Penzel, T., Pressman, M. and Iber, C. (2007). The visual scoring of sleep in adults. *Journal of Clinical Sleep Medicine* 3(2): 121–131. 233, 234

[30] Susmakova, K. (2004). Human sleep and sleep EEG. *Measurement Science Review* 4(2): 59–74. 233

[31] Tatum IV, W.O., Husain, A.M. Benbadis, S.R. and Kaplan, P.W. (2008). *Handbook of EEG Interpretation*. Demos Medical Publishing: New York. 240

[32] Traub, R., Bibbig, A., LeBeau, F.E.N., Buhl, E. and Whittingon, M. (2004). Cellular mechanisms of neuronal population oscillations in the hippocampus *in vitro*. *Annual Review of Neuroscience* 27: 247–278. DOI: 10.1146/annurev.neuro.27.070203.144303 235

[33] Vertes, R. (2005). Hippocampal theta rhythm: A tag for short-term memory. *Hippocampus* 15: 923–935. DOI: 10.1002/hipo.20118 235

[34] Vigh, J., Solessio, E., Morgans, C.W. and Lasater, E.M. (2003). Ionic mechanisms mediating oscillatory membrane potentials in wide-field retinal amacrine cells. *Journal of Neurophysiology* 90: 431–443. DOI: 10.1152/jn.00092.2003 235, 246

[35] Wang, X.-J. (1993). Ionic basis for intrinsic 40 Hz neuronal oscillations. *NeuroReport* 5: 221–224. DOI: 10.1097/00001756-199312000-00008 237

[36] Wang, X.-J. (2010). Neurophysiological and computational principles of cortical rhythms in cognition. *Physiological Reviews* 90: 1195–1268. DOI: 10.1152/physrev.00035.2008 237, 239

[37] Webb, W.B. (1986). Enhanced slow sleep in extended sleep. *Electroencephalography and Clinical Neurophysiology* 64: 27–30. DOI: 10.1016/0013-4694(86)90040-4 235

[38] Webb, W.B. and Dube, M.G. (1981). Temporal characteristics of sleep. In Aschoff, J. (ed.) *Handbook of Behavioral Neurobiology 4: Biological Rhythms*. Plenum Press: New York, pp. 499–522. 234

[39] Ziskind, E. (1965). An explanation of mental symptoms found in acute sensory deprivation: Research 1958–1963. *The American Journal of Psychiatry* 121: 939-946. 228

[40] Zubek, J.P. (1964). The effects of prolonged sensory and perceptual deprivation. *British Medical Bulletin* 20: 38–42. 228

A more recent review of the origins of the egg can be found in:
Buzaski, G. Anastassiou, C., and Koch, C. (2012). The origin of extracellular fields and currents—EEG, ECog, LFP, and spikes. *Nature Reviews—Neuroscience* **13** June 2012, pp. 407–420.

APPENDIX A

Modeling Approaches

A.1 MODELING ENDOCRINE SYSTEM USING MATLAB BY KEVIN FREEDMAN

Building a model of a biological system is inherently flawed due to the enormous complexity of even the simplest biological processes. The complexities arise due to the high integration of multiple subsystems that all interact and affect each other. Using models, however, the behavior of a simplified system can be analyzed and generalized to the real system with important assumptions. If the model behaved exactly like the system no matter what conditions or parameters are used, it would be the system and not just a model of the system.

The endocrine-hormone system was chosen as a model system that incorporates common control system components such as short- and long-range negative feedback, setpoints, and gain adjustments. The assumptions and simplifications of the actual endocrine-hormone model will be discussed here. One of the main assumptions is that any change in concentration of molecules become instantaneously distributed such that the effects of that molecules (interactions with other molecules or surrounding tissue) begin right away without any time delay. In the real system, a molecule is released from a cell or organ (the source) and the molecule has to diffuse or move by convection (through the bloodstream) in order to be dispersed within a biological fluid (i.e., cytosol or bloodstream). This is done by essentially removing the volume from the system and working only with the net amount of a certain biological molecule. In a volume-based system (which is not used here), a certain quantity of a molecule is released and in order to find the concentration, the amount of molecule (milligrams) must be divided by the volume of that fluid compartment (liters). In the present model, however, the relative amount of each molecule is found using a system of equations and the concentration is directly proportional to that amount by assuming that all fluid compartments in the body are held at a constant volume (which is generally a good assumption even though blood pressure/volume is a rhythmic circadian variable).

The systems of equations used in this model are based on two basic types of equations: the Michaelis-Menten enzymekinetics equation and the first-order chemical equation. Each biological molecule is mathematically connected to one or more other biological molecules. There is a general pathway of molecular interactions including receptor binding (assumed to follow first-order kinetics), signal transduction, enzyme binding, protein secretion, and other tissue specific binding. Each molecule can be thought of as a dynamic function of other molecules in the system. The dynamics of each molecule (given by the system of differential equations) is the most important part of this specific model since biological rhythms are generated at the molecular level. Therefore, the accuracy of this model in predicting actual in-vivo concentrations of system components was not a design goal. That being the case, the constants in each equation were arbitrarily chosen and were all of the same order of magnitude to prevent anomalous results (results that are highly dependent on specific and potentially unrealistic parameters). All the models were tested over a range of parameters to test the reproducibility of the models dynamics. If the characteristics of the model's response were unstable with respect to the models parameters, that particular simulation was not used. The mathematical formulations of the model and other assumptions are elaborated on below.

MODEL CHARACTERISTICS AND FURTHER ASSUMPTIONS

- Target Tissue Output is given a set point defined by the modeler.

  - When the target tissue output deviates from the set point, negative feedback acts on the internal stimulus and endocrine gland.

- The strength of the negative feedback is proportional to the concentration of the molecule exerting the negative feedback.

- The endocrine gland has a baseline level of secretion of hormone.

- Enzymes and carrier protein can saturate. Enzymes can only produce enzyme product as fast as V_{max} and carrier protein-hormone complexes can only reach a concentration equal to the initial starting concentration of carrier protein.

- The rate of change of enzymes and total carrier protein was zero (they were held at a constant concentration) unless it was desired that they oscillate, in which case a sinusoid term was given as its rate of change. Since enzymes are usually kept at a constant level inside cells (through negative feedback on gene transcription), this is a reasonable assumption.

- Cytoplasmic molecules and molecules in the bloodstream were assumed to degrade following first-order kinetics with the exception of enzymes and carrier protein as mentioned above.

- Molecules in the bloodstream are also broken down and excreted by the liver and kidneys and is modeled as an enzymatic reaction (there exists a maximum rate of excretion).

- If a rate equation did not exist, the ideal concentration was calculated and a rate equation was formulated to make the modeled concentration go towards the ideal concentration (if the modeled concentration was too high, a negative rate of change was imposed to decrease the modeled concentration; if the modeled concentration was too low, a positive rate of change was imposed to increase the modeled concentration).

- Receptor binding was modeled as a first-order process.

- Carrier protein binding of hormone is a reversible process. If hormone levels drop, the carrier protein releases the bound hormone in its active form.

MODELED EQUATIONS

$$\frac{d[P]}{dt} = k[ES] = k[E]\frac{[S]}{K_m + [S]} \qquad \text{Michaelis-Menten Equation}$$

where $k[E]$ is $V_{\max}$

$$\frac{d[P]}{dt} = k_{\text{forward}}[R] \qquad \text{First-Order Reaction}$$

VARIABLE LEGEND

C1	External Stimulus
C2	Internal Stimulus
C3	Intra-Endocrine Signal
C4	Enzyme for making Hormone
C5	Enzyme Product
C6	Hormone
C7	Carrier Protein
C8	Carrier Protein-Hormone Complex
C9	Target Tissue Output
C10	Liver Enzyme

VARIABLES CALCULATED AT EACH TIME POINT

Ideal Carrier Protein Saturation (ICPS) = (C7*C6) / (Constant1+C6)

Effective Internal Stimulus (EIS) = Set point - C1

REACTION RATE (RR) EQUATIONS

RR1 = Constant2*(EIS-C2)-Constant3*(C9-Set point)

RR2 = Constant4*C2-Constant5*(C9-Set point)

RR3 = Constant6*C4*C3/ (Constant7+C3)

RR4 = Constant8*C5-Constant9*C6

RR5 = Constant10*(ICPS-C8)

RR6 = Constant11*C10*C6/ (Constant12+C6)

RR7 = Constant13*C6

RR8 =Constant14*C10*C9/ (Constant 15+C9)

FINAL RATE EQUATIONS FOR EACH VARIABLE

$$dC1 = 0$$
$$dC2 = RR1$$
$$dC3 = RR2\text{-}kdeg*C3$$
$$dC4 = 0$$
$$dC5 = RR3\text{-}RR4\text{-}kdeg*C5$$
$$dC6 = BaselineH + RR4\text{-}RR5\text{-}RR6\text{-}RR7\text{-}kdeg*C6$$
$$dC7 = 0$$
$$dC8 = RR5$$
$$dC9 = RR7\text{-}RR8\text{-}kdeg*C9$$
$$dC10 = 0$$

A.2   MODELING NEURONS BY RAJARSHI GANGULY AND GEORGE NEUSCH

The simulations provided in the text modeling neuron activity were based upon another MatLab model, this one originally developed by Murat Saglam and available through the MatLab Central web site:

http://www.mathworks.com/matlabcentral/fileexchange/19669

The MatLab model is based upon the Hodgkin-Huxley Model of neuron function, as described in Figure A.1

The students were able to use the MatLab simulation to examine the effects of varying parameters, such as the external stimulus (I_ext), resting potential (V0), Nerst potentials (E_K, E_{Na}, etc.), and ion conductances on neuron output.

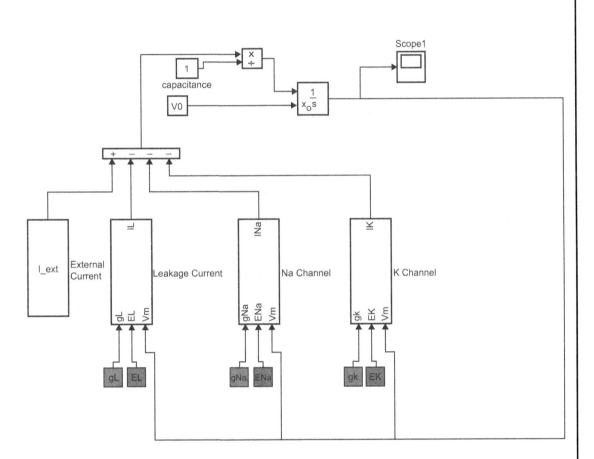

Figure A.1: Hodgkin-Huxley Model of neuron function.

APPENDIX B

The End of the Beginning

Life is all about timing... the unreachable becomes reachable, the unavailable become available, the unattainable... attainable. Have the patience, wait it out. It's all about timing.

— Stacey Charter

We have come to the end of Volume 1 of *Chronobioengineering*. What has been learned and what remains to be accomplished?

We started this journey with a single over-riding objective—to prove that biological rhythms are ubiquitous and fundamental characteristics of life on Earth. This proposition was supported by two assertions. First, complex goal-directed systems require precise timing to organize their activities in a coherent manner. Second, the magnitude of effects generated by geophysical cycles is so great that organisms on Earth must use rhythms to adapt to time-varying ecosystems. As we investigated these assertions, other reasons for the evolution of biological rhythms were uncovered. The nature of biological control systems, and of control systems in general, is inherently oscillatory. Many biological output functions—the pumping of blood, breathing, locomotion—are themselves oscillatory and require some form on rhythmic control mechanisms. Finally, physical constraints, such as those enforced on neural transmission, impose oscillatory solutions on systems that might not otherwise be rhythmic. Utilizing a variety of investigative tools—arguments, conceptual and mathematical models, observation and experimentation—we examined our fundamental proposition and established its validity. Biological rhythms really are ubiquitous and fundamental characteristics of all living systems.

From the point of view of a scientist, this might be sufficient in and of itself. To expand one's understanding, to test and support hypotheses, is the goal of scientific investigations and it appears we have accomplished much in furthering the understanding that rhythms are essential features of biological processes on every scale ranging from molecular to ecological. In fact, we were able to show how rhythms cascade throughout inter-connected networks such that oscillations at one level will permeate throughout life at all levels. Oscillate anywhere in the network of living systems and oscillations appear everywhere.

From an engineering point of view, the results of our investigation so far are not sufficient. Certainly, having established the importance of rhythmic phenomena in biology has implications for product and process design but these implications are not entirely clear as yet. There are some obvious repercussions based upon our discussion of circadian rhythms in terms of symptoms, treatment and detection but such daily rhythms are not the only cycles of importance. How can our understanding of neural oscillators be put to productive use and what about the other rhythms, such as pulsatile hormone secretion, central pattern generators, reproductive cycles, and even annual rhythms? What characterizes these other cycles and how can the knowledge of this multitude of oscillations and oscillators be utilized to improve human health, well-being, and productivity? To answer those questions is the goal of Volume 2.

Authors' Biographies

DONALD L. MCEACHRON

Dr. Donald L. McEachron is a Research Professor and Senior Lecturer and currently serves as the Coordinator for Academic Assessment and Quality Improvement for the School of Biomedical Engineering, Science and Health Systems at Drexel University. He holds a B.A. in Behavioral Genetics from the University of California at Berkeley and a Ph.D. in Neuroscience from the University of California at San Diego. In December 2006, Dr. McEachron received an M.S. in Information Science from Drexel University's iSchool. Dr. McEachron has worked extensively in the areas of imaging, editing three monographs on imaging applications in biomedicine, as well as numerous papers and presentations. However, Dr. McEachron's primary biomedical research has focused on chronobiology, biological rhythms, and human performance engineering. Dr. McEachron is presently developing programs to investigate the impact of light and circadian manipulations on human productivity, health, and well-being. In association with architects, civil engineers, and other engineers and social scientists, Dr. McEachron is working on the field of Indoor Ecology, examining how built environments influence human physiology and behavior. Dr. McEachron is also interested in investigating the impact of light and circadian manipulations on immune function in an animal model of AIDS. In addition, however, Dr. McEachron has published in a variety of other disciplines, including hominid evolution and education. He has served as PI or Co-PI on a variety of grants from both NIH and NSF involving autoradiographic image processing, neuroendocrinology and education. Dr. McEachron is currently involved in the design and implementation of computer-assisted knowledge management systems to augment instruction and assist in the development of personalized educational approaches, originally funded by the National Science Foundation and is assisting in the development of specialized lighting systems for nursing facilities under a grant from the Green Building Alliance. In addition to his work at Drexel, Dr. McEachron serves as Chair of the Engineering in Biology and Medicine Society, Philadelphia Chapter, IEEE Philadelphia section. In 2005 and again in 2012, Dr. McEachron was trained as an IDEAL Scholar in assessment practices by ABET, Inc.

BAHRAD SOKHANSANJ

Bahrad Sokhansanj grew up in Saskatoon, Canada and studied Engineering Physics at the University of Saskatchewan. He received a Ph.D. in Applied Science from the University of California, Davis, while doing thesis research in computational biophysics at Lawrence Livermore National Laboratory. After continuing on as a postdoctoral scientist at LLNL, he moved to Philadelphia,

PA and joined the School of Biomedical Engineering, Science and Health Systems at Drexel University. While in academia, his research and teaching interests included mathematical modeling of biological systems, with a focus on DNA repair and inflammation, as well as in the emerging field of metagenomics. He has recently changed careers, and after graduating, again, with a J.D. from Columbia University now works in technology-intensive patent litigation.

Printed in the United States
by Baker & Taylor Publisher Services